PETERSON'S 2000

AP* SUCCESS

BIOLOGY

Dana Freeman

Gordon Chenery

John Nishan

Peterson's
Thomson Learning™

Australia • Canada • Denmark • Japan • Mexico • New Zealand • Philippines
Puerto Rico • Singapore • Spain • United Kingdom • United States

Visit Peterson's Education Center on the Internet (World Wide Web) at
www.petersons.com

Editorial Development: American BookWorks Corporation
Contributing Editor: Barbara Maynard
Art and Design: James Snyder
Editorial Consultant: Barbara Maloney, Princeton Day School

ISBN 0-7689-0368-8

Printed in the United States of America

10 9 8 7 6 5 4 3 2 1

CONTENTS

INTRODUCTION

ABOUT THIS BOOK

The AP Biology exam will be a measure of how much you have learned throughout the year in your AP Biology class. This book has been written to help you achieve success in this exam, offered by the College Entrance Examination Board. It covers all the material offered on the AP Biology Examination, as well as some additional material on Laboratory exercises and pointers on how to write the exam's essay portion—the "free response" section.

There are three basic sections of this book. The first part, following the introductory material, presents a diagnostic exam. It is not as extensive as the actual exams but will test your knowledge of the material and will give you a good idea of how well you would do if you were to take the actual test now. You will also be able to determine what topics will require additional study. For the best results, take this test under simulated test-taking conditions. Find a quiet place in which to work and continue through the entire test without taking a break. It would also be helpful to time yourself. The purpose of keeping track of your time is to help you determine how quickly you can answer the questions. Evaluating both the results of the exam in terms of the number of questions you answered correctly, as well as the time it took to complete it, should help you determine how prepared you are for the final exam. Later on, when you take the full-length practice tests, you will be able to compare your progress—both in what you've learned and how quickly you were able to finish the test.

The second part of this book contains review material, presented in an outline format. Keep in mind that this is NOT a textbook, but rather an approach to reviewing for the test prior to taking the exam. By presenting the material in an outline format we are replicating class notes, and you can fill in additional information as you progress. In addition, we've included numerous illustrations to help you better understand the material. We've covered all the material that will be required by the test, and in addition, even given you a chance to write an essay, so that the free response questions in Section II of the exam will seem easier the day of the real test. As an additional "bonus," there is a vocabulary list at the end of each chapter. You should make the effort to learn their meanings.

The final part of this book offers you three, full-length simulated AP Biology exams. As you did with the diagnostic test, you should take these under simulated test-taking conditions. And of course, you should time yourself. Since the actual exam will take about 3 hours, make sure you set aside enough time to complete each practice test. When you have finished the first exam, take a break, and then go back to check your answers. Since this will take a considerable amount of time, perhaps you can wait until the following day to do this. Take the time and make the effort to find out what you marked wrong, and why. Go back through this book, check your textbook, or ask your teacher. You don't want to find a similar question on the actual exam and still not understand it.

Continue this same process on separate days for the final two exams. At this point, you should be able to achieve higher scores on each subsequent test, as well as improve upon the time it takes to finish each exam. Keep a record of both your scores and the time it takes to complete each exam to be able to measure your progress. Perhaps you can make a chart, like the one below.

EXAM	Section I Correct	Time to Complete	Section II Correct	Time to Complete
Diagnostic Test	(60 Q)/ _____)	(45 min)/ _____	(2 Q)/ _____)	(45 min)/ _____
Practice Test #1	(120 Q)/ _____	(90 min)/ _____	(4 Q)/ _____	(90 min)/ _____
Practice Test #2	(120 Q)/ _____	(90 min)/ _____	(4 Q)/ _____	(90 min)/ _____
Practice Test #3	(120 Q)/ _____	(90 min)/ _____	(4 Q)/ _____	(90 min)/ _____

When you've completed the entire chart, you will be able to see your progress, not only from test to test, but from the very beginning.

RED ALERT

ABOUT THE TESTS

As you prepare to take the AP Biology Exam, you are probably very aware of what subjects will be included on the test. However, for the sake of review, we will give you some general information about what to expect.

The AP Biology Exam consists of two sections. The first section consists of 120 multiple-choice questions on a variety of topics. This section represents about 60 percent of your total grade for this exam. The second section is the free response portion and requires you to write an essay for each of the four questions presented. This section is worth 40 percent of the total grade. In keeping with the format of this book, the answers to the free response question we provide are in outline format and present all of the information that can be included in your answer. Thus, when you check your answers against those given in this book, you should evaluate how much of your answer consists of the material we have provided. The more you have, the more complete your answer will be. You will be given one hour and 30 minutes to complete each section.

The exam covers a wide range of material, all of which you should have covered in your classroom. But to make it somewhat easier to focus on what you should know, following is an approximation of the percentage of each of the major topics that will be covered on the test itself.

Molecules and Cells — 25%
Chemistry of Life — 7%
Cells — 10%
Cellular Energetics — 8%
Heredity and Evolution — 25%
Heredity — 8%
Molecular Genetics — 9%
Evolutionary Biology — 8%
Organisms and Population — 50%
Diversity of Organisms — 8%
Structure and Function of Plants and Animals — 32%
Ecology — 10%

It should be readily apparent where you should focus your energies. For instance, if you're weak on Structure and Functions of Plants and Animals, you could be missing out on almost 1/3 of the test.

SOME GENERAL GUIDELINES

1. Work steadily but quickly. You will have approximately 45 seconds in which to answer each multiple-choice question. It doesn't leave a lot of time. And you have about 22 minutes for each essay.

2. Outline your essay before you write. It will make writing the essay easier. Keep in mind that an answer written in outline form is not acceptable. If you study the outlines we have presented in our sample tests, you will get an idea of what you can do to help formulate your thoughts.

3. You will be penalized for incorrect answers, 1/4 point per answer (or one point for every four incorrect answer). Therefore, guessing has a penalty. However, if you can narrow it down to only two choices, you stand a better chance of getting a correct answer.

4. To narrow down your choices, use the process of elimination. There are often several choices that are obviously incorrect, irrelevant to the question, or ones that just don't sound right. By eliminating those choices, you'll have a better chance of making an educated guess for the right answer.

5. Go for the essay questions first. They'll help build your confidence and eliminate any jitters if you can breeze through a lot of questions first. Skip the ones you don't immediately know. If you can save a few minutes here and there throughout the first section, you'll have more time in which to go back and work on those you didn't understand at first glance.

6. Keep track of the answer bubble. Too often, students who skip questions forget to skip the answer bubbles, and suddenly find that they've answered everything out of order. Make sure the answer numbers correspond to the questions.

7. Learn your vocabulary. The words that appear at the end of every chapter will be helpful on the test. Learn them, memorize them, and use them to associate words on the questions. If you see a word that you recognize, it will help trigger your memory and will enable you to work through the problem faster. It will also give you additional vocabulary that will be helpful when writing essays.

8. Draw clearly labeled diagrams. Diagrams can help illustrate your answers in the free response section. If the person who is

scoring your test can interpret your drawing, and the labels are accurate, it will help improve your score.

It's time to get to work. It's been proven that one of the major keys to success on standardized test taking is repetition. The more you practice, the better you will do on the real exam. It's straightforward—with practice come familiarity, and with familiarity comes understanding. Take the diagnostic test, review what you don't know, then take the practice tests. By the time you take the actual exam, you'll be ready. Good Luck!

AP BIOLOGY STUDY PLAN

As the AP Biology test gets closer, it's important to begin reviewing for the test as soon as possible. How much time do you have? If you have enough time to study comfortably, you'll be okay. If, however, you're pressed for time, you'll need to plan your studying carefully. We offer you these different study plans to help maximize your time and studying. The first is a 9-Week Plan, which involves concentrated studying and a focus on the sample test results. The second is the 18-Week Plan, or Semester Plan, favored by schools, and finally, The Panic Plan, for those of you who have only a few weeks to prepare. Obviously, the more time you have to prepare, the easier it will be to review all of the material and find yourself somewhat more relaxed when taking the actual exam.

These plans are supposed to be flexible, and only suggestions. You can feel free to modify them to suit your needs and your own study habits. But start immediately. The more you study and review the questions, the better your results will be.

THE 9-WEEK PLAN—2 LESSONS PER WEEK

Week 1

Lesson 1: Diagnostic Test. The Diagnostic AP Biology Test is designed to help you figure out what you need to know. Take this test under simulated test conditions, in a quiet room, and keep track of the time it takes to complete the test. The full exam consists of two parts—120 multiple choice questions and a free response (essay) section consisting of four questions. You are given about 3 hours in which to complete the exam. This test, however, is half that length.

Lesson 2: Diagnostic Test Answers. Once you have completed the test, spend this lesson carefully checking all of your answers, and reading through the explanations. This may take quite a bit of time, as will all of the tests, but it will enable you to select those subject areas that you should focus on, and the areas to spend the most amount of time studying.

Week 2

Lesson 1, Chapter One: Atoms and Compounds. Take your time to read through the first chapter. Note that the style of the review material is in an outline format. It should be similar to your classroom notes. Underline or use a marker to highlight those areas that are unclear to you.

Lesson 2, Chapter Two: Organization and Regulation. Again, read through this chapter, mark whatever is unclear, and go back and reread the material again, if necessary.

Week 3

Lesson 1, Chapter Three: Energy for Life: Anabolism and Catabolism. As you continue your lessons, try to study in a quiet room, uninterrupted by others in you household, or the TV, radio, or any outside noises.

Lesson 2, Chapter Four: Genetics: Mendel and Heredity. Again, read through this chapter, mark whatever is unclear, and go back and reread the material again, if necessary.

Week 4

Lesson 1, Chapter Five: Molecular Genetics: DNA and Evolution. You're approaching the half-way point in the content chapters of this book now, so continue reading and taking notes. You can, of course, break these lessons into sections. Work on half the chapter in the morning, and the other half in the afternoon.

Lesson 2, Chapter Six: Organisms and Populations. Read through this chapter, mark whatever is unclear, and then go back and reread the material again, if necessary. You can always ask your teacher for additional information if you're having difficulty.

Week 5   **Lesson 1, Chapter Seven: Animals: Structure and Function.** These are the last two chapters that cover the subject matter that will be on the test. By now you should have a strong understanding of the material.

Lesson 2, Chapter Eight: Behavior and Ecology. This is the final part of your journey into the subject review of AP Biology. All that's left are two more chapters that will help you improve your scores on the actual exam.

Week 6   **Lesson 1, Chapter Nine: The Laboratory.** An integral part of the AP Biology examination will be your understanding of the Laboratory and the techniques involved. This chapter will cover most of what you have already practiced in your classroom.

Lesson 2, Chapter Ten: Writing an Essay. The free-response section of the exam represents half of your grade, and covers more that just your understanding of the material. It is important for you to understand how to express yourself clearly. Practice writing some of the essays in this chapter and then check your answers against the outline of the topics that should have been covered in whatever you have written.

Week 7   **Lesson 1: AP Biology Practice Test #1.** Take this test and answer all of the questions you can, and then guess at those you don't know. Circle those questions that you guessed at so that you can zero in on those specific answers, and so that you don't delude yourself into thinking that you really knew those answers in the first place.

Lesson 2: AP Biology Practice Test #1 Answers. Check all of your answers to both parts of the test.

Week 8   **Lesson 1: AP Biology Practice Test #2.** Take this test and answer all of the questions you can. By now you will have noticed your improvement from when you took the diagnostic test to now.

Lesson 2: AP Biology Practice Test #2 Answers. Check all of your answers to both parts of the test.

Week 9   **Lesson 1: AP Biology Test #3.** This is the final test. By now you will have answered 360 multiple choice questions throughout these 3 practice tests, as well as having written 12 separate essays for the free-response sections. Although there may be some new material on the actual test that you haven't encountered here, you should be very well prepared.

Lesson 2: Biology Practice Test #3 Answers. Check all of your answers to both parts of the test, and then relax!

THE 18-WEEK PLAN—1 LESSON PER WEEK

If you're lucky enough to have the extra time, the 18-Week Plan will enable you to better utilize your study time. You will now be able to spread out your plan into one lesson a week. This plan is ideal because you are not under any pressure and you can take more time to review the material in each of the chapters. You will also have enough time to double-check the answers to those questions that might have given you problems. Keep in mind that the basis for all test success is practice, practice, practice.

THE PANIC PLAN

While we hope you don't fall into this category, not everyone has the luxury of extra time to prepare for the AP Biology test. However, perhaps we can offer you a few helpful hints to get you through this period.

Read through the official AP Biology bulletin and this *AP Biology Success* book and memorize the directions. One way of saving time on this, or any, test, is to be familiar with the directions in order to maximize the time you have to work on the questions.

Read the introduction to this book. It will be helpful in preparing for the test, and give you an understanding of what you can expect on the exam, and how much time you will have to complete both sections of the test.

Take the diagnostic tests as well as the practice tests. Practice writing the essays in the Chapter 10.

Focus whatever time you have left on those specific areas of the test that gave you the most difficulty when you took the practice tests.

Whatever time you have before the exam, keep in mind that the more you practice, the better you will do on the final exam.

DIAGNOSTIC TEST

AP BIOLOGY

SECTION I
Time—45 Minutes
Number of Questions—60

1. Monotremes possess which of the following characteristics?
   1-lay shelled eggs
   2-have separate openings for feces, urine, and reproductive products
   3-fur
   4-functioning mammary glands
   5-compound eyes

   (A) 1 and 2
   (B) 1 and 5
   (C) 2 and 4
   (D) 1, 3, and 4
   (E) 3, 4, and 5

2. Some portions of chromatin remain condensed permanently, even during the uncoiling after division. This chromatin is

   (A) euchromatin
   (B) heterochromatin
   (C) chromatid
   (D) prochromatin
   (E) histone

3. When a eukaryotic gene is transcribed, the primary RNA transcript contains

   (A) sequences complementary to the entire gene, but prior to protein synthesis, the introns are cut out.
   (B) sequences complementary to the entire gene, but prior to protein synthesis, the exons are cut out
   (C) only the exons which will be used in translation
   (D) only the introns which will be used in translation
   (E) defective sequences which will be excised by DNA polymerase prior to translation

4. MHC complexes and antibodies identify cells by

   (A) intracellular receptors
   (B) chemically gated ion channels
   (C) enzymatic receptors
   (D) G-protein–linked receptors
   (E) surface markers

5. Which of the following characteristics is found in eukaryotes, but is lacking in prokaryotes?

(A) flagella made of a single fiber of flagellin

(B) multicellularity

(C) circular DNA

(D) cellular respiration enzymes bound to the cell membrane

(E) enzymes for photosynthesis bound to cell membrane

6. Which of the following plant structures is triploid (3n)?

(A) megaspore mother cell

(B) embryo

(C) endosperm

(D) pollen grain

(E) pollen mother cell

7. Organisms for which an S-shaped growth curve is characteristic are called k-strategists. Which of the following is also a characteristic of k-strategists?

(A) numerous offspring

(B) early age of first reproduction

(C) parental care

(D) large brood size

(E) short generations

8. Pharyngeal slits develop from the embryo into what structures in humans?

(A) cochlea

(B) Eustachian tube

(C) auditory canal

(D) semicircular canal

(E) nasal cavity

9. The process of cohesion is responsible for

(A) leaves pulling water up from their roots

(B) the diffusion of water through a semipermeable membrane

(C) the stability of temperature of organisms

(D) lakes not freezing from the top down

(E) cooling of body surfaces during evaporation

10. Rhizoids play a major role in which of the following functions?

(A) anchor plant to substrate

(B) absorb water

(C) photosynthesis

(D) anchor plants to substrate and absorb water

(E) anchor plants to substrate, absorb water, and photosynthesize

11. Guttation is the loss of water through the

(A) stomata

(B) cells at the ends of veins in leaves

(C) root hairs

(D) weak spots in the Casparian strip

(E) vascular cambium

12. The diagram below represents a food web.

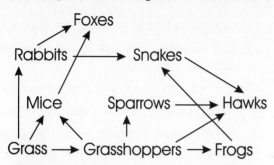

Which statement is supported by the information shown?

(A) Foxes, snakes, and sparrows are second-ary consumers.

(B) Snakes eat grass, grasshoppers, and frogs.

(C) Mice are carnivores.

(D) Rabbits, mice, and grasshoppers contain the greatest amount of stored energy.

(E) Sparrows and hawks are omnivores.

Questions 13 and 14 refer to the diagram shown below. It represents an experiment that was performed over a period of 12 hours.

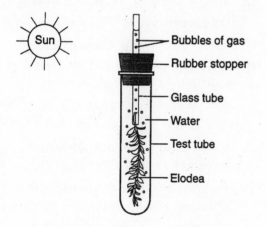

13. After the light source is changed from sunlight to pure green light, the rate of bubble production would most likely

(A) increase only

(B) decrease only

(C) increase, then decrease

(D) remain the same

(E) decrease for a short adjustment time, then dramatically increase

14. The triple bonds in atmospheric nitrogen are broken by

(A) nitrogen reductase

(B) ammonification

(C) denitrification

(D) catalase

(E) nitrifying bacteria

15. The diagram below represents some tissue as seen through a compound light microscope.

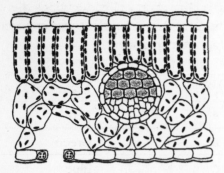

Which procedure was most likely used to prepare this tissue for viewing with the microscope?

(A) A small slice of a root tip was mounted on a slide.

(B) A thin cross section of a leaf was mounted on a slide.

(C) A cross section of Hydra was made, and a small piece of one side of the Hydra was mounted on a slide.

(D) A vertical cut was made through the body of an earthworm, and a thin slice was mounted on a slide.

(E) A section of the exoskeleton of a grasshopper was removed and mounted on a slide.

16. A centromere is a structure that joins the sister chromatids together during mitosis and is the site to which the kinetochores are attached. What comprises the centromere?

(A) Microtubules

(B) Microfilaments

(C) About 200 nucleotides

(D) 75–85 triglyceride molecules

(E) the steroid, centromosol

17. The diagrams below represent the stages in ecological succession. They are NOT in order. Which sequence represents a correct order of succession that would involve these stages?

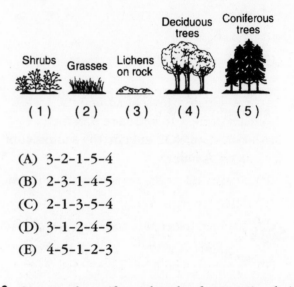

(A) 3-2-1-5-4

(B) 2-3-1-4-5

(C) 2-1-3-5-4

(D) 3-1-2-4-5

(E) 4-5-1-2-3

18. Structural motifs are bends of a protein chain that permit the protein to interlock with the major groove of the DNA helix. Which description fits the leucine zipper?

(A) shaped like a "Y," with the arms of the "Y" fitting into the major groove of the DNA

(B) two alpha helical segments linked by a short non-helical segment

(C) two hydrophilic protein chains, which bind to the minor groove of the DNA

(D) A zinc atom links an alpha-helical segment to a beta-sheet segment.

(E) An alpha-helix bends around, using hydrogen bonding to attach to the major groove.

19. The hastening of fruit ripening and the control of abscission of leaves, flowers, and fruits is due to which of these hormones?

    (A) ethylene

    (B) cytokinins

    (C) auxin

    (D) abscisic acid

    (E) gibberellin

20. In the diagram below, as an impulse travels from structure 1 to structure 5, which sequence does **NOT** include the secretion of a neurotransmitter?

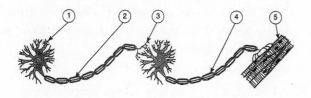

    (A) 1→4

    (B) 3→5

    (C) 3→4

    (D) 1→2

    (E) 2→4

21. Which of the following is present in a roundworm but absent in a flatworm?

    (A) coelomic cavity

    (B) digestive cavity

    (C) endoderm

    (D) coelomic cavity and digestive cavity

    (E) coelomic cavity, digestive cavity, and endoderm

22. Four plants of the same size and type were grown for 24 hours under identical conditions, except for the color of the light source. The chart below identifies the color of the light that each plant was exposed to during the experiment.

Plant	Light Color
A	Red
B	Green
C	Blue
D	White

After 24 hours, the leaves of each plant were tested for sugar. Which plant most likely contained the *smallest* amount of sugar?

    (A) red

    (B) green

    (C) blue

    (D) white

    (E) All plants would test for approximately the same amount of sugar.

23. When identical chemical groups are bonded to different atoms in two molecules, such as glucose and galactose, these are examples of

    (A) monosaccharides

    (B) polymer

    (C) monomers

    (D) structural isomer

    (E) stereoisomer

24. The process of ecdysis is a characteristic of

    (A) angiosperms

    (B) flatworms

    (C) ascomycetes

    (D) amphibians

    (E) arthropods

25. Vesicles derived from the Golgi apparatus that contain hydrolytic digestive enzymes are the

    (A) ribosomes
    (B) lysosomes
    (C) mitochondria
    (D) chloroplasts
    (E) microbodies

26. The diagrams below show the general appearance of five tree fruits that were used in an experiment to determine the length of time necessary for each type of fruit to fall a set distance. One hundred fruits of each type were used, and the average time of fall for each type of fruit is shown in the data table below.

Fruits

Silver maple   Norway maple   White ash   Red oak   Shagbark hickory

(Not drawn to scale)

Tree Type	Average Fall Time of 100 Fruits (seconds)
Silver Maple	3.2
Norway Maple	4.9
White Ash	1.5
Red Oak	0.8
Shagbark Hickory	0.8

What could a student correctly infer about the distribution of these fruits if they fell from branches 5 meters above the ground while the wind was blowing at 20 miles per hour?

(A) A silver maple fruit would land closer to the base of its parent tree than would a shagbark hickory fruit.

(B) A white ash fruit would land farther from the base of its parent tree than would a silver maple fruit.

(C) A white ash fruit would land closer to the base of its parent tree than would a shagbark hickory fruit.

(D) A Norway maple would land farther from the base of its parent tree than would a silver maple fruit.

(E) The red oak and shagbark hickory fruits would land the farthest from the base of their parent trees.

27. The reaction below is a hypothetical enzymatic pathway:

$$A \rightarrow B \rightarrow C \rightarrow D$$
$$+ \rightarrow F$$
$$E$$

If F is an allosteric inhibitor and binds to molecule B, which of the following is likely to happen?

(A) The amount of C will build up as F is produced.

(B) The amount of A will decrease when F is produced.

(C) As E is produced, the amount of B will increase.

(D) As D and E react, the amount of F will decrease.

(E) As F is produced, the amount of D will decrease.

28. The biome characterized by relatively long, cold winters, soil which is relatively fertile and herds of grazing mammals is

(A) savanna

(B) temperate grassland

(C) taiga

(D) temperate deciduous forest

(E) tundra

29. The function of the stigma in Euglena is

(A) storage of starch

(B) attachment of the flagella

(C) photosynthesis

(D) phototropism

(E) water regulation

30. Which of the following is an underground stem?

(A) mycorrhiza

(B) rhizoid

(C) rhizome

(D) rhodopsin

(E) rhabdom

31. An organism which is eukaryotic, unicellular, and heterotrophic would most likely be placed in which of the following groups?

(A) Monera

(B) Bacteria

(C) Protista

(D) Fungi

(E) Plantae

Questions 32 and 33 are based on the diagram of the nitrogen cycle below. In the diagram, letters A through E represent organisms carrying on a process at that particular point in the cycle.

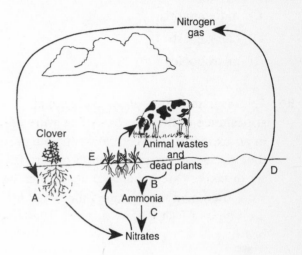

32. Letter B represents

(A) scavengers

(B) decomposers

(C) autotrophs

(D) synthesizers

(E) carnivores

33. Nitrifying bacteria are represented by letter

(A) A

(B) B

(C) C

(D) D

(E) E

34. Small circular auxiliary genomes, which have the capability to enter and leave the main genome, are

(A) transposons

(B) plasmids

(C) homologous chromosomes

(D) oncogenes

(E) satellite DNA

35. The diagram below represents root epidermal cells.

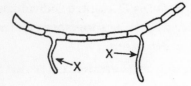

The major function of the structures labeled X is to

(A) transport carbon dioxide from the roots to the leaves

(B) move food particles into the phloem

(C) absorb complex food particles for transport to the leaves

(D) store food for the plant in the form of starch

(E) increase the surface area for absorption of water

36. A madreporite, stone canal, ampullae, and ossicles are structures found in

(A) bony fish

(B) cartilaginous fish

(C) coral

(D) echinoderms

(E) crustaceans

37. All of the following isolating mechanisms are prezygotic mechanisms **EXCEPT**

(A) species reproducing during different seasons

(B) hybrids are sterile

(C) little or no attraction between species

(D) gamete fusion does not occur due to chemical factors

(E) species occur in different areas

38. The diagram below represents the change that occurred after a fluid was added to a wet mount of some *Elodea* leaf cells.

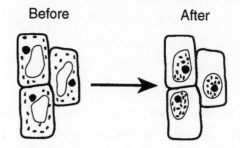

Before   After

Which fluid(s) was most likely added to the wet mount?

(A) salt solution

(B) tap water

(C) distilled water

(D) iodine solution

(E) tap water or distilled water

39. All of the following are mutualistic relationships except which one?

(A) fungi and plant roots in mycorrhizae

(B) alga and fungus in lichens

(C) ants and aphids

(D) tapeworm and intestine

(E) ants and the plant *Acacia*

40. The major factors in the path of *Drosophila* development are given. Which answer places them in their proper order?

1-imaginal discs
2-larval instars
3-maternal genes
4-metamorphosis
5-syncytial blastoderm

(A) 3, 2, 1, 5, 4

(B) 1, 5, 4, 2, 3

(C) 3, 4, 1, 2, 5

(D) 3, 5, 2, 1, 4

(E) 1, 4, 3, 5, 2

41. A major process in the production of ammonia in the body is through

(A) hydrolysis of polysaccharides

(B) deamination of amino acids

(C) beta-oxidation of fatty acids

(D) glycolysis of sugars

(E) replication of DNA

42. In humans, sperm are produced by spermatogenesis and ova are produced by oogenesis. If both processes occur normally, what is the resultant number of each gamete formed?

(A) One sperm and one ovum

(B) Two sperm and two ova

(C) Four sperm and four ova

(D) Four sperm and one ovum

(E) One sperm and four ova

43. Which of the following relationships would be symbiotic?

(A) commensalism only

(B) mutualism only

(C) parasitism only

(D) commensalism and mutualism

(E) commensalism, mutualism, and parasitism

44. What is the result of self-pollinating plants that normally outcross?

(A) It produces more uniform offspring.

(B) It produces ill-adapted individuals because it brings together deleterious recessive genes.

(C) It produces a greater number of offspring.

(D) It produces a greater number of mutations.

(E) It produces fertile offspring.

45. Which of the following fungal structures is diploid?

(A) zygospore

(B) ascospore

(C) conidium

(D) badiospore

(E) primary mycelium

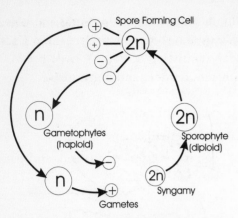

Spore Forming Cell

Gametophytes
(haploid)

Sporophyte
(diploid)

Gametes

Syngamy

46. This life cycle would be characteristic of which of the following organisms?

(A) bacteria

(B) green algae

(C) mushrooms

(D) ferns

(E) mammals

47. Three babies, X, Y, and Z, were not identified prior to going to the hospital nursery. Baby X had blood type A, Baby Y had blood type AB, and Baby Z had blood type O. The three sets of parents, K, L, and M, were AB and A, AB and O, and B and O, respectively. This information is summarized below:

Baby	Blood Type	Parents	Blood type
X	A	K	AB, A
Y	AB	L	AB, O
Z	O	M	B, O

Which answer correctly matches the baby with its parents?

(A) X with K, Y with L, Z with M

(B) X with L, Y with K, Z with M

(C) X with M, Y with L, Z with K

(D) X with L, Y with M, Z with K

(E) X with K, Y with M, Z with L

48. All of the following will disrupt the Hardy-Weinberg gene equilibrium except which one?

(A) immigration

(B) selection

(C) random mating

(D) genetic drift

(E) mutation

Questions 49 and 50 refer to the diagrams of molecules below.

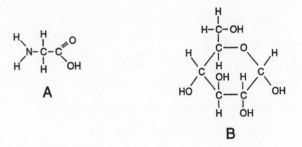

A

B

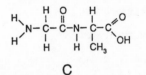

C

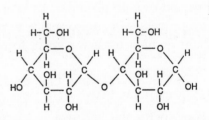

D

E

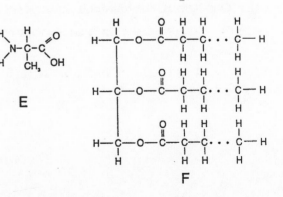

F

Peterson's AP Success: Biology

49. Which food contains the highest percentage of the type of molecule represented by F?

    (A) bread

    (B) lettuce

    (C) table sugar

    (D) butter

    (E) potato

50. Amino groups are found in which diagrams?

    (A) A, C, and E

    (B) B and D

    (C) A, C, E, and F

    (D) F only

    (E) B, D, and F

51. Which of the following appendages is analogous to a bat wing?

    (A) forelimb of a frog

    (B) eagle wing

    (C) housefly wing

    (D) whale front fin

    (E) human arm

52. In C_4 plants, the majority of RuBP carboxylase is bound to ____ because _____.

    (A) CO_2; CO_2 is present in high concentrations

    (B) CO_2; CO_2 is produced in the mesophyll cells

    (C) O_2; O_2 is present in high concentrations

    (D) O_2; CO_2 is present in high concentrations

    (E) O_2; O_2 is produced in the mesophyll cells

The diagram below represents the digestive system of a student who has eaten a sandwich consisting of two slices of bread, chicken, lettuce, and mayonnaise.

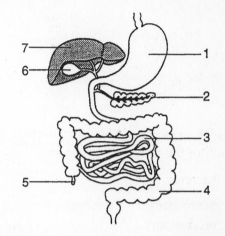

53. The final reactions for the complete hydrolysis of the bread would occur in organ

    (A) 1

    (B) 2

    (C) 3

    (D) 4

    (E) 5

54. In the Five Kingdom system of classification, which of the following Kingdoms is characterized by its members being eukaryotic, lacking chlorophyll, lacking motile cells, and having a cell wall made of chitin or other non-cellulose polysaccharide?

    (A) Monera

    (B) Protista

    (C) Fungi

    (D) Plantae

    (E) Animalia

55. Prosimians were the first primates and were successful for two reasons. One reason was that they possessed grasping fingers and toes. The other was

    (A) homeothermy
    (B) ability to walk on two legs
    (C) monogamous mating
    (D) binocular vision
    (E) omnivorous diet

56. Which of the following best describes the DNA molecule?

    (A) Two parallel stands of nitrogen bases held together by hydrogen bonding.
    (B) Two complementary strands of deoxyribose and phosphates held together by hydrogen bonding.
    (C) Two antiparallel strands of nucleotides held together by hydrogen bonding.
    (D) A single strand of nitrogen bases coiled upon itself by hydrogen bonding.
    (E) A single strand of nucleotides coiled into a helix.

57. The stimulus of light of the correct wavelength and intensity, a series of cold days, or the passage of time at an appropriate temperature, are all signals for

    (A) flower production
    (B) pollen formation
    (C) fruit ripening
    (D) the end of seed dormancy
    (E) pollination

58. Why don't the endonucleases found in bacteria cleave their own DNA?

    (A) The pH in the bacteria is too low to activate enough molecules of the endonucleases to cleave the DNA.
    (B) The temperature in the bacterial cell is too warm and is above the optimum temperature for endonucleases, so the enzyme activity is greatly decreased.
    (C) There are no nucleotide sequences in the bacterial chromosome recognized by the endonuclease.
    (D) Bacterial DNA do not show the two-fold rotational symmetry recognized by endonucleases.
    (E) Bacteria modify their own DNA by adding methyl groups to some of the nucleotides within their recognition sequence.

59. Which of the following characteristics accounts for decreased fluidity in the cell membrane?

    (A) unsaturated phospholipid molecules and lateral switching of the phospholipid molecules
    (B) saturated phospholipid molecules with the presence of cholesterol
    (C) unsaturated phospholipid molecules with lateral switching
    (D) rotation with switching sides
    (E) lateral switching with increased cholesterol

60. Duchenne's muscular dystrophy is a disease which results in the wasting away of muscles. It is caused by a sex-linked recessive gene. A female is a carrier and a male is normal. Their first child, a female, has a normal genotype (and is not a carrier). What are the chances that their second child will also have a normal genotype?

(A) 0 percent

(B) 12.5 percent

(C) 25 percent

(D) 50 percent

(E) 75 percent

SECTION II
Time–45 Minutes
Number of Questions–2

1. The enzyme catalase catalyzes the decomposition of hydrogen peroxide into water and oxygen as shown by the reaction below:

$$\text{catalase}$$
$$H_2O_2 \rightarrow H_2O + O$$

The production of water and oxygen is affected in the reaction by several variables: the pH of the surroundings, the temperature at which the reaction occurs, and the amount of substrate present.

(A) Pick any ONE of these three factors and design a controlled experiment that shows how the variable affects the production of water and oxygen.

(B) Explain how you would measure the rate of the reaction.

(C) Describe the results you would expect. Explain why you would expect these results.

2. Structure relates to function in many places in various organisms. Choose THREE of the following examples and for each example, describe how the structure of that part of the organism relates to its function relating to *water flow*.

(A) leaf epidermis

(B) cell membrane

(C) nephron

(D) xylem

ANSWERS AND EXPLANATIONS

QUICK-SCORE ANSWERS					
1. D	11. B	21. A	31. C	41. B	51. C
2. B	12. A	22. B	32. B	42. D	52. A
3. A	13. B	23. E	33. C	43. E	53. C
4. E	14. A	24. E	34. B	44. B	54. C
5. B	15. B	25. B	35. E	45. A	55. D
6. C	16. C	26. D	36. D	46. D	56. C
7. C	17. A	27. E	37. B	47. B	57. D
8. B	18. A	28. B	38. A	48. C	58. E
9. A	19. A	29. D	39. D	49. D	59. B
10. A	20. D	30. C	40. D	50. A	60. D

SECTION I

1. **The correct answer is (D).** Monotremes lay eggs with shells and have fur and functioning mammary glands. They only have one opening for feces, urine, and reproductive products and do not have compound eyes, which is an arthropod characteristic.

2. **The correct answer is (B).** Heterochromatin is the coiled form of DNA and protein. Some chromatin remains coiled at all times, and the DNA of this chromatin never gets expressed. Euchromatin is the remaining part of the chromatin, which does get expressed.

3. **The correct answer is (A).** Introns are long sequences of nucleotides, which do not code for any portion of the polypeptide. The transcript contains the sequences that are complementary to the entire gene; however, prior to protein synthesis, the introns are removed. This process is called RNA splicing.

4. **The correct answer is (E).** Surface markers are proteins or glycolipids in the cell membrane. Intracellular receptors have no extracellular signal-binding sites. Chemically gated ion channels are transmembrane proteins forming a pore. Enzymatic receptors are single-pass transmembrane proteins and G-protein–linked

receptors are transmembrane proteins with a cytoplasmic binding site, such as found in the rod cells of the eyes.

5. **The correct answer is (B).** Prokaryotes lack true multicellularity. They may aggregate in clumps or colonies, but there is no communication between cells. All of the other characteristics are prokaryote characteristics.

6. **The correct answer is (C).** The endosperm, which nourishes the developing embryo of angiosperms, is formed from the union of a male nucleus and two polar nuclei of the embryo sac and is therefore triploid (3n). The megaspore mother cell, embryo, and pollen mother cell are all diploid (2n), while the pollen grain is haploid (n).

7. **The correct answer is (C).** All of the other characteristics are for r-strategists, which, in addition, have no parental care.

8. **The correct answer is (B).** The Eustachian tube runs from the middle ear to the throat and is formed from the pharyngeal slit found in the human embryo.

9. **The correct answer is (A).** Movement of water up from the roots is due to the hydrogen bonding holding water molecules together. Choice (B) is due to their high polarity. Choice (C) is due to the high specific heat of water. Choice (D) is due to the density of ice decreasing after 4°C. Choice (E) is due to the high heat of vaporization of water.

10. **The correct answer is (A).** While rhizoids are rootlike structures, the amount of water that they absorb for the plant is minimal. The great major of water is absorbed directly through the leaves. Their primary function is to anchor the plant to the substrate.

11. **The correct answer is (B).** Guttation is due to excess water pressure in a plant. The water is forced out through specialized cells, which are at the ends of veins in the leaves.

12. **The correct answer is (A).** All three of these animals eat other animals, making them secondary consumers.

13. **The correct answer is (B).** Plants are unable to absorb green light for use in photosynthesis. The bubbles being produced are oxygen, which, along with sugar, are the products of photosynthesis. By changing the light source to green, this light will be reflected, and the rate of photosynthesis will decrease, thereby producing fewer bubbles.

14. **The correct answer is (A).** This is the enzyme that breaks apart the diatomic nitrogen molecule. Ammonification is the process of releasing nitrogen in the form of ammonium ions through the decomposition of proteins by bacteria and fungi. Denitrification

is the conversion of nitrates in the soil into elemental nitrogen and dinitrogen monoxide by certain bacteria. Nitrifying bacteria aid in converting amino acids into nitrates in the soil. Catalase is the enzyme that aids in the breakdown of hydrogen peroxide into water and oxygen.

15. **The correct answer is (B).** The key to identifying this is the open area on the underside of the specimen, which is a stoma. The two small cells at either side of the opening are the guard cells, which open to allow entry of carbon dioxide and close to prevent water loss.

16. **The correct answer is (C).** The centromere is made of approximately 220 nucleotides, made of highly repeating DNA sequences. Microtubules are hollow protein cylinders found in places such as centrioles, flagella, and cilia. Microfilaments are part of the cytoskeleton of the cell. Choice (D) is incorrect, and choice (E) is a nonsense answer.

17. **The correct answer is (A).** Lichens are pioneer vegetation, followed by grasses, which lack any woody tissue. Small shrubs will follow them. Then coniferous trees, followed by deciduous trees.

18. **The correct answer is (A).** Choice (B) describes a helix-turn-helix, and choice (D) describes a zinc finger. Choices (C) and (E) don't describe any kind of structural motif.

19. **The correct answer is (A).** Ethylene causes all of the above affects on plants.

20. **The correct answer is (D).** A neurotransmitter is secreted, which the impulse passes over a synapse, the junction between two neurons or between a neuron and an effector cell. The synapse is located between the axon of the first neuron, which ends in structure 3 and the dendrite of the second neuron. All of the pathways given in the question require the impulse to pass over the synapse except choice (D), which only has the impulse traveling along the first neuron from the cell body through the axon.

21. **The correct answer is (A).** Both flatworms and roundworms have a digestive cavity and endoderm, but the flatworms are acoelomate, while the roundworms have a pseudocoelom.

22. **The correct answer is (B).** Plants reflect light of green wavelength. Consequently, they would be unable to convert this color into usable light for photosynthesis and would be unable to make sugar.

23. **The correct answer is (E).** This is the definition of stereoisomers. Choice (A): While glucose and galactose are both monosac-

charides or simple sugars, this only pertains to the number of carbons and the ratio of carbons to hydrogens to oxygens. Choice (B): A polymer is a large molecule made of identical smaller units, called monomers. Choice (D): A structural isomer has the same formula but has the functional group located off a different carbon atom.

24. **The correct answer is (E).** Ecdysis is molting, and of the above groups, only the arthropods molt to allow their bodies to grow.

25. **The correct answer is (B).** Lysosomes are derived from the Golgi and contain hydrolytic enzymes. While microbodies also contain hydrolytic enzymes, they are not derived from the Golgi apparatus but from other microbodies. Ribosomes function in protein synthesis, mitochondria in energy production, and chloroplasts are involved in photosynthesis.

26. **The correct answer is (D).** The Norway maple, having an average fall time of 4.9 seconds, should land the farthest from the base of its parent tree.

27. **The correct answer is (E).** As F is produced, being an allosteric inhibitor, it will bind to B and slow down the reaction, thereby decreasing the amount of D that is produced in the reaction. C is consumed in the production of F. E is not part of the reaction flow, so it will be unaffected. As D and E react, the quantity of F will increase, because D and E are making it.

28. **The correct answer is (B).** These characteristics are all indicative of the temperate grassland.

29. **The correct answer is (D).** The stigma is light sensitive. It allows the Euglena to move toward light, where it can photosynthesize to make food.

30. **The correct answer is (C).** Mycorrhiza is the symbiotic relationship between a fungus and the roots of a plant. A rhizoid is a slender projection from mosses or liverworts, which grow into the substrate and serves to anchor the organism. Rhodopsin is the photopigment in rods in the eye. Rhabdom is the light-sensitive central core of an ommatidium, a visual unit of a compound eye.

31. **The correct answer is (C).** Protista members fit this description. Examples are Paramecium and Amoeba. Both Monera and the older phylum Bacteria are prokaryotic. Most fungi, choice (D), are multicellular although the yeasts are unicellular. However, this is an exception to this group. Plantae members are autotrophic.

32. **The correct answer is (B).** Decomposers break down dead material and return the constituents to the soil.

33. **The correct answer is (C).** Nitrifying bacteria convert ammonia into nitrates, which are usable by plants.

34. **The correct answer is (B).** Transposons are DNA sequences that have the ability to move around the DNA but are not circular. Homologous chromosomes are pairs of chromosomes that carry alleles for the same traits. Oncogenes are cancer-causing genes. Satellite DNA is a nontranscribed region of the chromosome.

35. **The correct answer is (E).** The root epidermal cells, also referred to as root hairs, greatly increase the surface area that a root has for absorption of water into the plant.

36. **The correct answer is (D).** Echinoderms possess these structures, which are part of their water vascular system.

37. **The correct answer is (B).** This is a postzygotic mechanism. Any organism that is formed by it is infertile. All of the other answers are prezygotic mechanisms. They occur before the sperm fused with the egg.

38. **The correct answer is (A).** In the "After" diagram, water left the cells due to the environment, thus becoming hypertonic to the cell. There are more dissolved substances in the environment, which causes water to leave the cell, in an attempt to reestablish equilibrium between the cell and its environment. The only one of the solutions added to the wet mount that would be hypertonic to the cell is the salt solution.

39. **The correct answer is (D).** In all of the other relationships, both members benefit from the relationship. The tapeworm is a parasite of vertebrate intestines, taking nutrients from the host and harming the host, not aiding it in any way.

40. **The correct answer is (D).** The first factor in the development is the maternal genes. The syncytial blastoderm is a structure made of one cytoplasm with about 4,000 nuclei and is formed after fertilization. There are three larval instars to allow the larva to grow in size. Imaginal disks, about twelve groups of cells, are present in the abdomen of the larva. They have no role in the development of the larva but will form key parts of the adult. Metamorphosis involves first changing the larva into a pupa and then the pupa into an adult.

41. **The correct answer is (B).** The deamination of amino acids results in the formation of NH3. Hydrolysis of polysaccharides produces simpler disaccharides. The beta-oxidation of fatty acids produces acetyl Co-A. Glycolysis of sugars produces pyruvate. Replication of DNA produces two new DNA molecules, which are identical to the original.

42. The correct answer is (D). Spermatogenesis results in the production of four equal sperm, all of which are viable. However, oogenesis, during the first meiotic division, produces one secondary oocyte and a first polar body. This is due to the unequal distribution of cytoplasm, with the ovum receiving most of the cytoplasm during this first division. During the second meiotic division, the primary oocyte divides into a secondary oocyte, which will become the ovum, and a secondary polar body. The first polar body may divide again into two smaller polar bodies. Thus, the primary oocyte will produce one ovum and one to three polar bodies.

43. The correct answer is (E). A symbiotic relationship is where two organisms live together. All three of these relationships involve two organisms living together.

44. The correct answer is (B). If plants that outcross contain deleterious genes, which are recessive, these genes are more likely to remain masked due to normal dominant genes. When these plants self-pollinate, the incidence of homozygous recessive plants with the deleterious genes is increased. The plants with this genetic condition are generally less suited for survival.

45. The correct answer is (A). The zygospore is formed by the union of two haploid cells and therefore is diploid. All of the other structures result from meiotic division and are haploid.

46. The correct answer is (D). Sporic meiosis is characteristic of green plants.

47. The correct answer is (B). Baby X goes with Parents L, Baby Y goes with Parents K, and Baby Z goes with Parents M. Parents K could only have a child with blood types AB, A, or B. Parents L could only have a child with blood types A, B, or O, and Parents M could only have a baby with blood types B or O. Since Parents K are the only ones who could have an AB type baby, then Y is theirs. Since Parents M cannot have a baby with blood type A, this baby must belong to Parents K. That leaves Baby Z to have Parents M.

48. The correct answer is (C). All of the other factors will cause the equilibrium to shift since they are pressures outside the normal situation. Only random mating will not disrupt the equilibrium once it has been established.

49. The correct answer is (D). Butter is high in fats, and figure F is a triglyceride, which is a type of lipid.

50. The correct answer is (A). The amino group is $-NH_3$, which is found only in these three molecules. They are all amino acids.

51. **The correct answer is (C).** Analogous structures are those which have a similar function but form different embryologically. The other four answers are homologous structures. They have different functions from a bat wing, but their embryological origins are the same.

52. **The correct answer is (A).** Even though both O_2 and CO_2 can bind to RuBP carboxylase, C_4 plants use quite a bit of energy to concentrate CO_2 within the cells that carry out the Calvin Cycle. Thus, due to this high concentration, most of the available enzyme binds to the CO_2.

53. **The correct answer is (C).** Final carbohydrate breakdown occurs in the small intestine. The stomach, (1), is responsible for the beginning of protein digestion. The pancreas, (2), secretes pancreatic juice into the small intestine, but is an accessory organ and food does not enter it. The colon, or large intestine, (4), does not carry out any digestive activity and neither does the appendix, (5).

54. **The correct answer is (C).** Fungi have all of the stated characteristics. Monera are prokaryotic, Protista have motile cells and some have chlorophyll, obviously Plantae possess chlorophyll and have a cell wall of cellulose, and Animalia lack any cell wall and have motile cells.

55. **The correct answer is (D).** This allowed them to judge distances more accurately than other tree-dwelling animals with which they were in competition for food. The combination of grasping fingers and toes and binocular vision was an advantage to these animals.

56. **The correct answer is (C).** DNA is composed of two strands of nucleotides, which run in opposite directions. This is referred to as antiparallel. Nucleotides consist of a 5-carbon sugar, a nitrogenous base, and a phosphate group. The nitrogenous bases are held together by hydrogen bonding.

57. **The correct answer is (D).** All of these factors will signal a seed to end the dormancy period. Water alone is not enough to necessarily end the dormancy period of a seed.

58. **The correct answer is (E).** By adding methyl groups to some of the nucleotides within their recognition sequence, the bacteria make the endonuclease nonfunctional.

59. **The correct answer is (B).** If the fatty acid tails are saturated, that is, filled to capacity with hydrogen molecules or having cholesterol present, the membrane decreases in fluidity. The other characteristics all contribute to increased fluidity.

60. The correct answer is (D). The chance of the second child being normal has nothing to do with the condition of the first child. Each child has the same chance of getting the disease or of being as normal as any other child. Since the mother is a carrier, she can either pass a normal allele or a muscular dystrophy allele to the child. The odds of passing either are 50 percent. The father is normal, so he can pass either a normal gene to a female child or no gene to a male child. Since the father can pass on only a normal X or a Y chromosome, inheritance is up to the mother. Half of her children, regardless of sex, will inherit a normal allele from her, and half will inherit the allele for the disease.

SECTION II

1. Part A: Each experimental characteristic may earn 1 point. A maximum of 7 points for this section. Only 3 points maximum if the experiment will not work.

- Statement of hypothesis for the experiment
- A control is specified, such as room temperature, pH of 7, or a stated amount of substrate
- Identify and hold constant at least one variable, such as time, amount of catalase added, etc.
- Manipulate the independent variable (change pH, change temperature, increase/decrease substrate amount)
- Describe what is being measured to determine rate (decrease in H_2O_2, formation of H_2O or release of O_2)
- Quantify the measurement of the variable (method and time frame of measurement)
- Give the rate calculation or a definition of rate
- Verify the results through sample size (>1) or repetition
- Utilize statistical application of data (mean, t-test, ANOVA, etc.)
- Designing an exemplary experiment

Part B: Describe expected experimental results (Maximum 2 points)

- Written or graphic description of expected experimental results (1 point)
- At low temperatures, enzyme activity is low
- At high temperatures, enzyme activity is low
- At a mid temperature, such as 35°C, enzyme activity is high

- At low (acidic) pH, enzyme activity is low
- At high (basic) pH, enzyme activity is slow
- At a neutral pH, the optimum for this enzyme, enzyme activity is high
- At low levels of substrate, enzyme activity is low
- Increasing the level of substrate increases the activity
- At a certain point, enzyme activity levels off and increasing the amount of substrate does not increase the activity.
- The graphs below represent 2-pt graphs, but to earn any points, graphs must be labeled entirely.

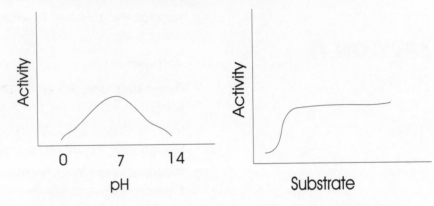

Part C: Biological explanation of results (maximum 2 points)

Temperature

- discussion of enzyme kinetics or metabolic changes
- enzyme denatures at temperatures above 40°C
- enzyme activity slows down at lower temperatures but does not denature pH
- discussion of change in the balance between positively and negatively charged amino acid residues
- loss of three-dimensional shape of enzyme preventing it from fitting into active site
- optimum pH where enzyme functions best

Amount of Substrate

- increasing amount of substrate increases the number of interactions between enzyme and substrate
- more opportunity for an enzyme to react with a substrate molecule if concentration is higher
- reaction reacts at a constant rate where all of the enzyme molecules are engaged in breaking down a substrate molecule.

2. Each answer would earn 2 points. (a maximum of 4 points for each of the three parts answered)

a. Leaf epidermis

- covered with waxy cuticle, which prevents water loss.
- contains stomata, which allow entry of CO_2 but also allow water to escape.
- contain veins, which transport water into the leaf for photosynthesis.
- intercellular spaces have a saturated atmosphere, which prevents the loss of water from the mesophyll cells
- selective permeability to solutes can influence direction of water flow.

b. cell membrane

- due to lipid bilayer, cell membranes are freely permeable to water.
- water molecule is small enough to pass through cell membrane
- contains porin channels, which are proteins, that allow free movement of water.

c. nephron

- Bowman's capsule at the beginning of the nephron absorbs water from the glomerulus.
- Countercurrent flow of material in the descending and ascending limbs of the loop of Henle creates a concentation gradient that allows water to be reabsorbed into the blood
- Nephron is lined with thin epithelial cells.
- Nephron is in close association with capillaries, allowing water to be reabsorbed easily into blood.

d. xylem

- consists of dead, hollow, tubular cells that are end to end
- tracheids in angiosperms move water from cell to cell through pits in the secondary wall.
- vessels connect end to end and move water efficiently
- tracheids have pit-pairs in their primary walls and move water less efficiently than vessels do.

Unit I—Molecules and Cells

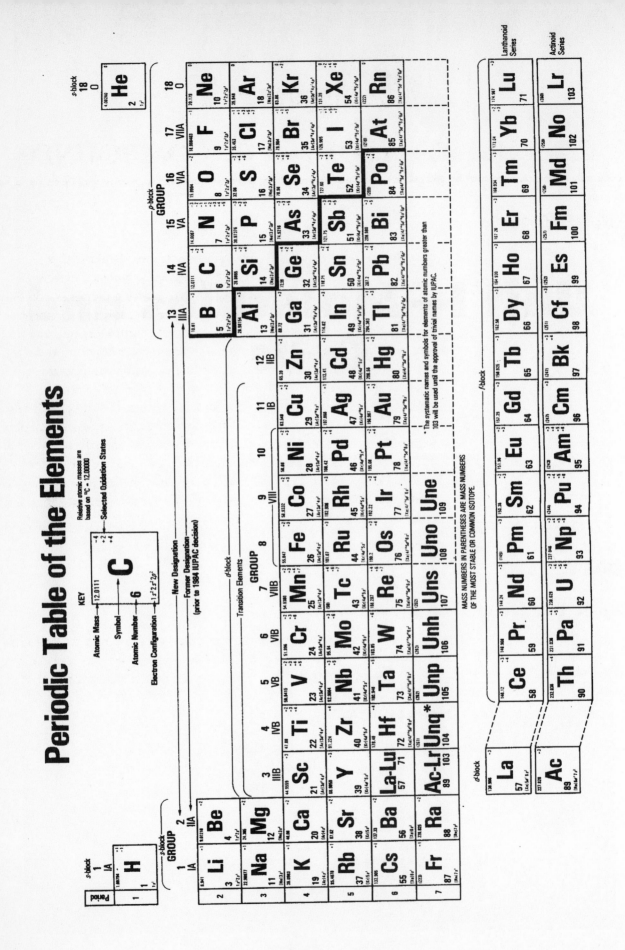

Periodic Table of the Elements

Chapter 1

CHEMISTRY: ATOMS AND COMPOUNDS

OVERVIEW

Appreciation for the **unity** of life on earth must begin with an understanding of the **diversity** of the building blocks, beginning at the atomic level. More than 90 naturally occurring elements combine in what seems to be an infinite number of ways to produce surprisingly similar classes of chemical compounds, both inorganic and organic. In addition to composition, these substances bond their elements together in a variety of ways. Organisms and organic molecules are complex. Complexity derives from using atoms, information, and energy. On the AP Biology test, the student should expect a detailed assessment of the properties, structures, formation, and breakdown of these substances. In addition, the student is expected to know the role and the effect of energy in the interactions between elements and compounds, particularly in the anabolism and catabolism of the chemicals of life. We begin, then, at the atomic level.

CHEMISTRY

ELEMENTS

Forms of matter that cannot be broken down into any simpler form by ordinary means. The more than 100 elements on the periodic chart are distinctly different from one another in discreet ways.

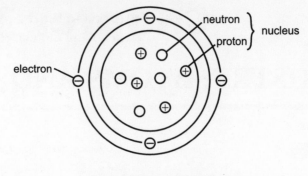

Typical atom (Beryllium)

Atoms

The smallest part of an element that is still that element and is composed of more than 40 subatomic particles. The three main subatomic particles are the proton, neutron, and electron.

A. Nucleus

Dense center of the atom where most of the mass is concentrated.

1. *Proton*—Positive particle in the nucleus of an atom; its mass is arbitrarily set as 1 for the purpose of calculating the three mass of an atom.

2. *Neutron*—Neutral particle in the nucleus of the atom, generally agreed to be a proton and an electron combined. Its mass number is also considered to be equal to 1 for the purpose of calculating the mass of an atom, since the electron is 1/1837 the mass of a proton and is, therefore, not significant in mass calculations.

B. Orbital(s)

1. *Electron*—Negative particle that remains some distance from the nucleus while still being a part of the atom.

2. *Atomic Models*

   - *Planetary*—Abandoned theory that had electrons orbiting the nucleus as the planets in a solar system orbit a central star. For the purpose of writing structural formulae of atoms and compounds, we usually use this model.

   - *Bohr*—Indicated that electrons were discreet packets of energy placed outside the nucleus, depending on their present quantum or energy state.

   - *Electron Cloud Model*—Current theory that states, as best we can determine, the location of the electron is a statistical statement, placing it somewhere in a "cloud" surrounding the nucleus, depending on its energy state.

C. Atomic Number

The number of protons in an atom.

D. Atomic Mass

The number of protons and neutrons in the nucleus of an atom. Electrons are not considered in this number. The actual number on the periodic table is an average of the abundance of all the isotopes of each element.

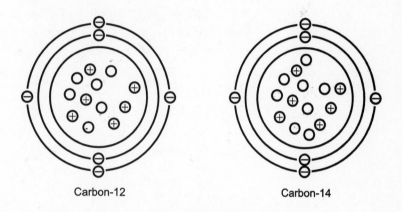

Carbon-12   Carbon-14

Isotopes of Carbon

E. Isotope

Different form of an atom due to a different number of neutrons. Carbon 12 has six neutrons, while Carbon 14 has eight neutrons. Different isotopes of an element have slightly different properties that are sometimes biologically important.

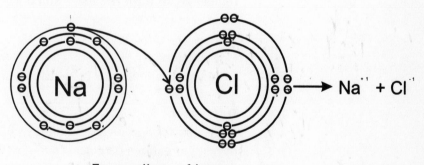

Formation of Ions

F. Ion

Electrically charged atom due to a loss (+ ion) or gain (ion) of electrons. An ionic bond is a "sticking together" of ions by virtue of the loss/gain of electrons that takes place between them. As atoms lose electrons and thus become positive, attraction occurs with the atoms that gain electrons and thus become negative.

Some essential atoms

A. Hydrogen
Simplest element that has one proton and one neutron. It combines with oxygen to form water, an essential abiotic substance for living things.

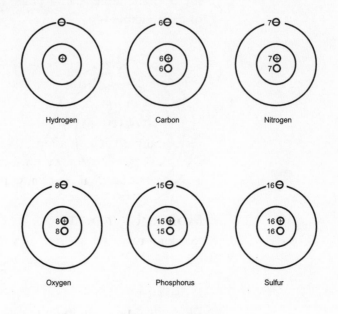

Atoms of Life

B. Carbon
Ubiquitous element on this planet and forms the basis for all living things. It can form the most number of bonds in chemical combinations.

C. Nitrogen
Essential component of proteins and nucleic acids.

D. Oxygen
Necessary for aerobic respiration to take place and in the formation of many organic molecules.

E. Phosphorous
Necessary for nucleic acid structure and in energy transformations. Also important in membrane structure.

Peterson's AP Success: Biology

F. Sulfur
Forms sulfide bridges that are found in proteins.

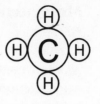

Methane molecule

MOLECULES

Group of chemically combined atoms that are the smallest form of a compound and has the properties of that compound.

Compounds
Chemically combined atoms that form discreet particles in which the atoms lose their individual physical properties and the compound, as a combined unit, takes on new properties.

Molecular bonds
"Sticking together" of atoms by sharing electrons.

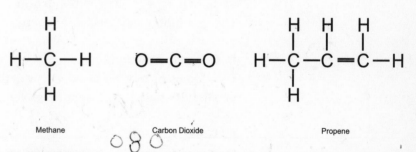

Methane   Carbon Dioxide   Propene

A. Covalent
Sharing of pairs of electrons between atoms that may result in single, double, or triple bonds. A single bond, for example, results from the sharing of one pair, thus the reference to "single."

1. *Nonpolar covalent*—Covalent sharing is distributed symmetrically within the molecule.

2. *Polar covalent*—Covalent sharing is distributed nonsymmetrically within the molecule, resulting in weak positive and negative charges.

B. Hydrogen bonds
Caused by polar covalent bonding, hydrogen bonds are the weak attractions between slightly positive hydrogen atoms of a molecule

and slightly negative atoms, such as oxygen and nitrogen, of another molecule.

Molecular forces

Forces that affect molecules, including the following:

A. Van der Waals
Attractive forces that occur between electrically neutral molecules because they are so close to each other.

B. Hydrophobic
Clumping together of molecules that are insoluble in water because they are nonpolar.

Some molecules essential to life

Carbon Dioxide   Oxygen   Water

A. Carbon Dioxide
Carbon and two oxygens bonded in a nonpolar, covalent orientation. This is a highly oxidized, low energy form of carbon.

B. Oxygen molecule
Two atoms of oxygen bonded covalently. Molecular oxygen allows a maximum oxidation of organic molecules in aerobic respiration.

C. Water
One oxygen and two hydrogen atoms bonded polar covalently. Supplies the hydrogen and electrons to initiate the photosynthesis process.

WATER

Essential abiotic molecule for sustaining life. It is essential in photosynthesis, in the maintenance of membranes, and as the solvent for all of life's molecules.

Structure
Two hydrogens with two single, covalent bonds with an oxygen atom forming a polar molecule.

Peterson's AP Success: Biology

A. Shape
Orientation of the unbonded electrons in the oxygen and the protons in the hydrogen orient them on "opposite sides of the molecule with respect to each other." The bond angles are approximately 105 degrees.

B. H+ bonds
Weak associations between the negative side of the water molecule oxygen and the positive side that the hydrogens create; bonds that hold water molecules together and result in high specific heat; good solvent properties and cohesion that results in a substantial surface tension.

Properties
Water has several unique properties.

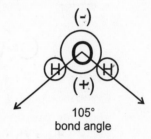

105°
bond angle

Water molecule

A. Polar
The covalent bond orientation results in the slightly positive hydrogens being on one side of the molecule and the unequally unshared electrons from oxygen more often on the other side of the molecule.

B. Cohesion
The dipole nature of the molecule produces bonds between the positive hydrogens and the negative oxygen that, in sum, can be substantial. The result is that water molecules tend to stick together, resulting in a very high surface tension.

C. High Specific Heat
Because of the hydrogen bonds, water absorbs a large amount of heat before it vaporizes.

D. Weak acid/base
Water has a neutral pH of 7.0 and normally dissociates into equal amounts of H+ ions and OH ions. It can, therefore, be a weak proton donor (an acid) or a weak proton acceptor (a base).

ACIDS/BASES

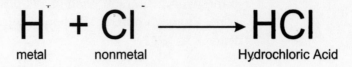

$$H^+ + Cl^- \longrightarrow HCl$$

metal   nonmetal   Hydrochloric Acid

Formation of an acid

Acids

Proton donor; when dissociated in aqueous solutions, acids become sources of protons donated to the reaction.

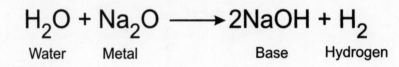

$$H_2O + Na_2O \longrightarrow 2NaOH + H_2$$

Water   Metal   Base   Hydrogen

Formation of a base

Bases

Proton acceptor; when dissociated in aqueous solutions, bases become proton acceptors in a reaction.

Peterson's AP Success: Biology

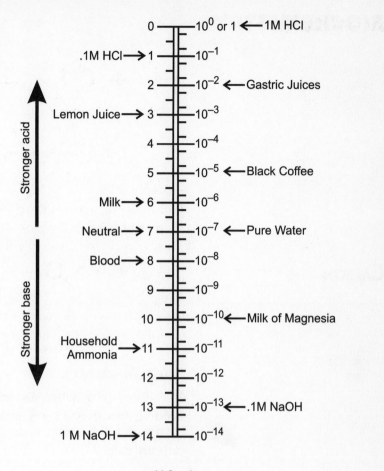

pH Scale

pH

Measure of concentration of hydrogen ions in a solution. The term stands for the *percent concentration of hydrogen ions* and is the negative log of the concentration of hydrogen ions. Thus a lower pH means a higher concentration of H+. (NOTE: hydrogen ions (H+) and protons are the same thing.) One may also take the inverse of this concentration and thereby arrive at the concentration of hydroxyl ions in solution.

BIOCHEMISTRY

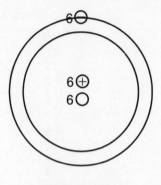

Carbon Atom

CARBON

Atom that is the basis for life on earth, owing to its versatility. Carbon has four bonding electrons that make it capable of forming many complex molecules. The chemistry of life is the chemistry of carbon.

Carbohydrates

Molecules that organisms create for energy storage or as structural components, such as cellulose. Their basis is carbon, hydrogen, and oxygen in a 1:2:1 ratio, thus the original term of hydrated carbon or carbohydrate.

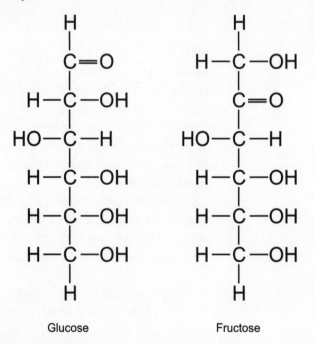

Glucose   Fructose

A. Monosaccharides

Simple sugars made up of a backbone of usually three to seven carbons with the formula Of $C_1H_2O_1$. When combined to produce

Peterson's AP Success: Biology

polysaccharides, they give off water at each bonding site in a reaction known as dehydration synthesis.

1. *Glucose*—Preferred energy molecule of life. It is produced in chloroplasts in the process of photosynthesis that stores energy from the sun in organic molecules and broken down in the process of aerobic respiration that releases energy.

2. *Fructose*—Fruit sugar with the same molecular formula as glucose but different structural formula. When combined with glucose in a dehydration synthesis reaction, the two form sucrose.

B. Disaccharide

Complex sugars formed from the union of two monosaccharides in a condensation reaction.

Condensation

Also called dehydration synthesis. When two monosaccharides bond together to form a disaccharide and a water molecule is produced, two hydrogens and one oxygen are removed from the two simple sugars and formed into a water molecule in the process. Complex organic molecules are created by assembling a small number of simple subunits in many different ways, analogous to forming many words from the 26 letters of the alphabet. The assembly of complex molecules from simple subunits is by so-called "dehydration synthesis" or "condensation reactions" (the same thing). When the small subunits are combined, a hydrogen atom (H) is removed from one and a hydroxyl group (OH) from the next. These atoms combine to make a water molecule (HOH), and the subunits become covalently bonded. The opposite process is called hydrolysis (water splitting) and releases the simple subunits from the molecule. This occurs, for example, during digestion.

Hydrolysis

Opposite of condensation, it occurs when carbohydrates, for example, are metabolized. Two hydrogens and one oxygen are ultimately inserted into the polysaccharide to cleave it into two simpler carbohydrates. For each cleavage, one water molecule is needed.

A. Polysaccharide

Carbohydrate formed from a number of simple sugars; the typical storage and structural form of carbohydrate in autotrophs. Plants have no preference. Polysaccharides are less diverse than DNA or proteins and are often long chains of a single monosaccharide, such as glucose (e.g., starch, cellulose, and glycogen).

1. *Starch*—Plant polysaccharide that is almost insoluble in water. Upon the formation of glucose in autotrophs, starch is quickly formed to conserve the glucose, which is soluble in water.

2. *Cellulose*—Largest carbohydrate molecule is insoluble in water and can be metabolized primarily by bacteria and fungi. An autotroph structural molecule, it is nonetheless important to heterotrophs.

3. *Glycogen*—The heterotroph analog to starch, it is used to store energy by animals.

Proteins

One of the main structural as well as functional molecules in living systems, protein comes from chains of amino acids bonded together. It is found embedded in plasma membranes, acting as catalyst, or as part of the structural integrity of many parts of living systems. In addition to the C, H, N, and O mix of living systems, they also contains S. Proteins are assembled from 20 different amino acids.

① Central C with Hydrogen

② R-group

③ Carboxyl

④ Amine group

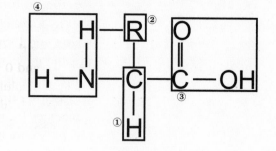

Amino acid

A. Amino acids

Main building block of polypeptides.

1. *Backbone*—Amino acids are built around a carbon center that has a hydrogen attached; one of the four covalent bonds of the central carbon.

2. *R-group*—Occupies a second of the four covalent bonds of the central carbon and can be as simple as a hydrogen or a complex ringed structure. There are 20 R-groups on the 20 amino acids that form proteins.

3. *Carboxyl group*—COOH complex attached to a third spot on the central carbon.

4. *Amine group*—NH complex attached to the fourth covalent bond on the central carbon.

B. Polypeptides

Built up of strings of two or more amino acids that are bonded together as a result of condensation.

C. Protein structures

Four different ways in which proteins are found in living things.

1. *Primary*—The number, type, and sequence of amino acids forming a linear structure.

2. *Secondary*—Repetition of amino acid sequences and, therefore, bond angles that gives the linear primary structure molecule a repeating structural pattern.

3. *Tertiary*—Folding pattern of the chain that relates to the underlying secondary structure and gives the molecule a three-dimensional shape.

4. *Quaternary*—Fitting together of two or more independently folded polypeptides that conform to each other to provide a functional protein.

Lipids

Molecules that store the most energy for living systems; also made up of C, H, and 0 that are not in a 1:2:1 ratio. Lipids are insoluble in water but soluble in nonpolar substances. Fats are examples that have three fatty acid residues attached to a glycerol backbone. Lipids are often assembled from fatty acids and an alcohol.

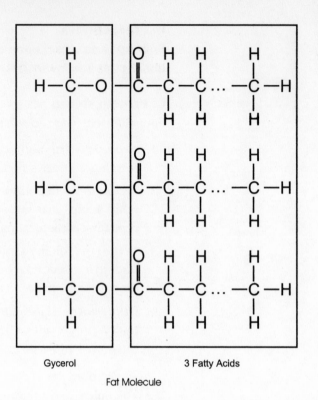

Gycerol   3 Fatty Acids

Fat Molecule

A. Fats

Three fatty acids bonded at their carboxyl site to the hydroxyl site in a molecule of glycerol and often referred to as a triglyceride. They are either saturated and contain all single covalent bonds between carbon atoms in the fatty acids or are unsaturated and contain a varying number of double covalent bonds between carbon atoms in the fatty acids by the removal of two hydrogens at that site.

B. Oils

Esters of open-chain hydrocarbons that are liquid at room temperature.

C. Waxes

Esters of long chain fatty acids and long chain alcohols. The have low melting points but are solids at room temperature.

D. Steroids

Fashioned on a backbone of four linked carbon rings and include cholesterol and hormones.

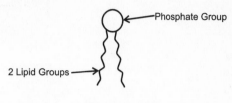

Phospolipid Molecule

E. Phospholipids

Composed of two fatty acids (the hydrophobic side of the molecule) and a phosphate group (the hydrophilic side of the molecule). Found in plasma membranes of cells as two layers of molecules (bilayer) where the hydrophobic ends are oriented to both the inside of the bilayer and the outsides of the bilayer.

Nucleic

Acids control molecules of life that contain or carry the codes for all the molecules that a particular cell needs. In addition, they carry the code for making new cells of a like kind—the key to heredity. Their building blocks are known as nucleotides and are composed of a 5-carbon sugar, a phosphate group, and a nitrogenous base. Their almost infinite variability lies in the unique sequences of their nucleotides.

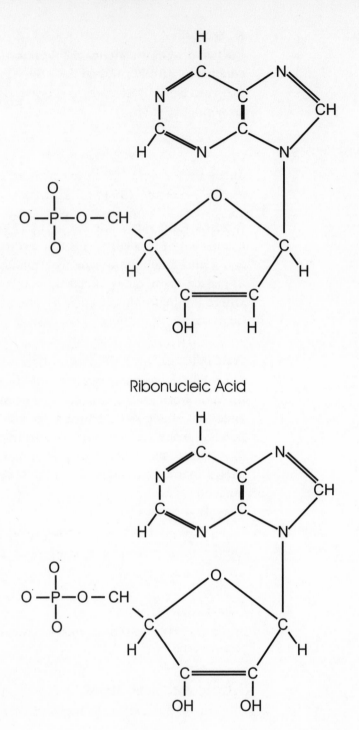

Ribonucleic Acid

Deoxyribonucleic Acid

A. DNA

Nucleic acid that contains the 5-carbon sugar known as deoxyribose, which denotes less oxygen than the ribose sugar. This molecule is the basis for genes and chromosomes. DNA is assembled from only four nucleotides, but the "words" are millions of nucleotides long.

B. RNA

Nucleic acid that contains the 5-carbon sugar known as ribose, which denotes more oxygen than the deoxyribose sugar. It is the molecule that carries the DNA code to the ribosome in cells and helps in fashioning proteins.

CHEMICAL REACTIONS

Interactions among chemicals fall into five categories and are characterized by an alteration of the structures involved as well as their energy levels. An organism that is properly nourished has the atoms/molecules it needs. The information for its complexity comes from DNA. The energy comes from manipulating oxidation/reduction reactions. Oxidation is the loss of electrons and, therefore, the loss of the energy of those electrons. In biological systems, this is usually the result of adding oxygen to a molecule or removing hydrogen. In cellular respiration, energy is removed from organic molecules by removing hydrogen atoms. Carbon is discarded in its most oxidized form, CO_2. Reduction is the opposite of oxidation. Thus, every chemical reaction oxidizes something and reduces something else. Reduction can be the loss of oxygen or the gain of hydrogen. Biological molecules are highly reduced and, therefore, have lots of hydrogens and are complex. The business of being alive is to maintain reduced molecules in an oxidizing environment. In other words, we eat to stay reduced.

Synthesis

Combining of two or more substances (atoms or molecules) that results in the making of another substance unlike the original components.

Decomposition

Breakdown of a substance into its component parts; it is the opposite of synthesis.

Single replacement

One part of a substance is replaced by another.

Double replacement

Similar to single replacement, except the exchange results in two new compounds as a result of the replacement in two substances, not one.

Equilibrium

Results when conditions produce no net changes in amounts of reactants and products. This can be altered by the addition of reactants, products, or any one of a number of physical aspects of the reaction, such as temperature, acidity, etc.

Free energy

Becomes available for work as a result of a reaction.

A. Exergonic

Release of energy during a reaction that is spontaneous.

B. Endergonic

Absorption of energy during a reaction that is nonspontaneous.

Enzyme

Organic catalysts (proteins) that alter the amount of energy needed to initiate a reaction, usually accelerating a reaction. They take part in the reaction but are not altered by it.

Coenzyme

Works with an enzyme to complete the catalytic reaction and consists of either a metal or nonprotein organic molecules. Vitamins can be coenzymes.

Co-factor

Any organic or inorganic substance that is needed for the operation of an enzyme.

MULTIPLE CHOICE QUESTIONS

1. In the reaction rate versus substrate concentration graph below, the curve plateaus because

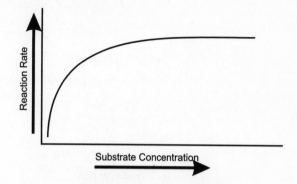

Energy curve for Reaction Rate versus Substrate concentration

   (A) a noncompetitive inhibitor is present

   (B) a competitive inhibitor is present

   (C) all the substrate has been converted to product

   (D) the active site is saturated with substrate

   (E) the cofactor is locked in an inactive conformation

2. Which of the following terms does NOT refer to an example of a weak force of interaction between two molecules?

   (A) covalent bond

   (B) Van der Waals bond

   (C) hydrophobic bond

   (D) electrostatic bond

   (E) hydrogen bond

3. The element found in all amino acids that is NOT found in carbohydrates is

   (A) sulfur

   (B) carbon

   (C) oxygen

   (D) hydrogen

   (E) nitrogen

4. Which of the following would NOT have hydrophilic properties?

   (A) OH

   (B) molecules in aqueous solution

   (C) long hydrocarbon chains

   (D) ionized molecules

   (E) polar molecules

5. Which if the following is NOT found in DNA?

   (A) uracil

   (B) thymine

   (C) phosphate

   (D) adenine

   (E) deoxyribose sugar

6. Elements used for making molecules found in living systems must

   (A) contain nitrogen and sulfur

   (B) be hydrophilic

   (C) be covalent

   (D) be water soluble and ionic

   (E) be suitable and available

7. Which of the following is NOT a property of water?

   (A) hydrophilic properties

   (B) hydrophobic properties

   (C) very good solvent

   (D) density as a solid is less than that as a liquid

   (E) strong surface tension

8. Which of the following are weak forces usually associated with biological molecules?

   (A) hydrogen bonds

   (B) electrostatic interactions

   (C) covalent bonds

   (D) ionic bonds

   (E) both (A) and (B)

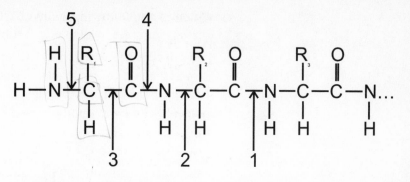

9. Which arrow(s) point to a peptide bond?

(A) 1

(B) 1 and 4

(C) 2

(D) 3

(E) 2 and 5

10. A polypeptide that is ten amino acid units long is split into several small fragments, and the sequences of some of the fragments are discovered. The fragments are lys-trp-arg, lys-asp-ala-gly, pro-gln-his-lys, and arg-pro-gln. What was the primary source of the polypeptide?

(A) ala-gly-ser-gln-lys-trp-arg-pro-gln-his

(B) asp-ala-gln-ser-gln-his-lys-trp-arg-pro

(C) ala-gly-pro-gln-his-lys-trp-arg-pro-asp

(D) lys-trp-arg-pro-gln-his-lys-asp-ala-gly

(E) gly-ala-asp-lys-his-gln-pro-arg-trp-lys

EXPLANATION OF ANSWERS FOR MULTIPLE CHOICE QUESTIONS

1. **The correct answer is (D).** A reaction in which the enzyme is the catalyst may be written as follows:

Enzyme + Substrate ↔ Enzyme-Substrate complex ↔ Enzyme + Product

The substrate binds to a specific site on the surface of the enzyme, known as the active site, after which product and enzyme are released. The enzyme is then available to bind another substrate. At low substrate concentration, the reaction rate increases sharply with increasing substrate concentration because there are abundant free enzyme molecules available to bind to an added substrate. At high substrate concentration, the reaction rate reaches a plateau as the enzyme active sites become saturated with substrate. The enzyme-substrate complex and no free enzymes are available to bind the added substrate.

2. **The correct answer is (A).** All of the choices except choice (A) are weak forces in biological systems. Covalent bonds involve a sharing of electrons, which leads to a strong interaction between molecules. In large numbers, some of the forces listed can produce strong binding in molecules, but taken individually, they cannot compare to the strength of a covalent bond.

3. **The correct answer is (E).** Amino acids derive their name from the amine group that bonds to one of the four covalent bonds of the central carbon atom. The amine group consists of one nitrogen and two hydrogens and is found in all molecules known as amino acids. The amine group forms a peptide bond with the COOH complex of an adjoining amino acid in the construction of a polypeptide.

4. **The correct answer is (C).** Any substance with electrical charges will exhibit hydrophilic or "water loving" properties and interact with water forming hydrogen bonds. Ionized molecules are charged molecules, as are polar molecules. The long hydro-carbon chains on molecules of that nature distribute any charge they might have uniformly over the chain, thus rendering the chain nonpolar and thus nonresponsive to the polar nature of water necessary for hydrophilic properties in the molecule.

5. **The correct answer is (A).** Choices (B) through (E) are all present in DNA. Thymine and adenine are two DNA bases and make up one of the three component of nucleic acids, namely nucleotides. These consist of a base, a 5-carbon sugar, and a phosphate group. The sugar in DNA lacks an oxygen and is called deoxyribose sugar. Uracil is unique to RNA, as is the 5-carbon ribose sugar not listed here.

6. **The correct answer is (E).** All of the choices listed are found in one or more biologically significant molecules, but choice (E) must be true of all biologically significant molecules. They must be correct for the "job," and they must be within reach of the living system that needs them. Not all molecules contain nitrogen and sulfur, nor are they attracted to water. In addition, they are not all covalent substances.

7. **The correct answer is (B).** Choice (B) translates to "water fearing" and, therefore, is a logical impossibility since it would describe water as not binding with water. The hydrophobic as well as the hydrophilic nature of a molecule is used to describe the interaction another molecule has with water. More impor-tantly, it describes a situation that would not have a polar nature, which water does. Hydrophilic WOULD be explained by the

polar nature of water and refer to the fact that it attracts polar substances. Choices (C) through (E) also reflect interactions that result from the polar nature of water.

8. **The correct answer is (E).** The hydrogen bonds formed in biological molecules refers to the positive nature of available hydrogens, such as those found in the bases of nucleic acids, that promote the nucleotide base pairing. In addition, electrostatic forces are often associated with molecule-molecule binding, such as that found in enzyme substrate complexes. Finally, choices (C) and (D) represent large forces and, therefore, do not fit the nature of the question posed.

9. **The correct answer is (B).** A peptide bond, formed when two amino acids are joined, occurs at the site where the amino group—NH_2—of one amino acid and the carboxyl group—$COOH$—are bonded. One H^+ of the amino group is removed and coupled with an OH^- from the carboxyl group to form water, which is removed from the molecule, leaving two amino acids bonded together. Choice (B) indicates the arrows that show these bonding situations.

10. **The correct answer is (D).** Assembling the four fragments in the following way:

lystrparg
    argprogln
        proglnhislys
            lysaspalagly

lystrpargproglnhislysaspalagly

indicates that choice (D) is the correct answer.

VOCABULARY

acidic

adhesion

amino acids

atom

basic

carbohydrates

cellulose

chemical reaction

cohesion

condensation

compound

dehydration synthesis

deoxyribonucleic acid

dipeptide

disaccharides

electron

elements

fat

glucose

glycogen

hydrogen bonds

hydrolysis

hydrophilic

hydrophobic

lipids

monosaccharides

nucleic acids

nucleotides

nucleus

neutron

organic compounds

oxidation

peptide bond

phospholipids

pH scale

polypeptide

polysaccharides

protein

proton

reactants

reduction

ribonucleic acid

specific heat

starch

synthesis

Chapter 2

CELLS: ORGANIZATION AND REGULATION

OVERVIEW

Cells are the basic unit of living things. The student should demonstrate a knowledge of the structure and function of cells as well as a detailed knowledge of the cell organelles. This area includes the way that cells divide and produce the next generation of cells. The student should be able to compare and contrast mitosis and meiosis with ease, including the phases of the cell cycle. A brief discussion of the regulatory mechanisms involved in the life of the cell is included. Chapter 3 in this unit details the necessary components of the two principal energy-related processes: photosynthesis and aerobic respiration. Unit II, Chapters 1 and 2, discuss heredity and protein synthesis in more detail. To begin with, we define the cell and discuss the cell organelles.

CELL TYPES

The principle types of cells on earth are the prokaryotic and eukaryotic. Viruses are not considered living things by many people.

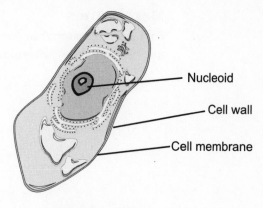

Prokaryotic Cell

Nucleoid
Cell wall
Cell membrane

PROKARYOTIC

Lack the membrane-bound organelles found in eukaryotic cells and consist of the Archaea, bacteria, and the blue-green algae. They generally have a plasma membrane, a single DNA molecule not coupled with protein molecules, small ribosomes, cytoplasm, and a cell wall. If they possess flagella, they are different from the eukaryotic flagella and cilia. There are two very different kinds of prokaryotic cells, those called bacterial (including cyanobacteria or blue-green algae) and those called Archaea. The Archaea are in many ways more like eukaryotes than bacteria. Many Archaea can tolerate extremes of temperature, pressure, pH, and salinity.

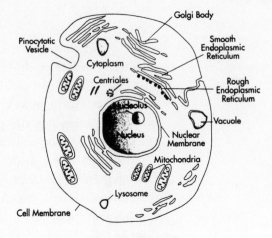

Eukaryotic cell

Eukaryotic

Contain some or all the organelles listed below and are found in or compose all of the non-prokaryotic living things. Evolutionarily, they are "younger" than the prokaryotes.

Membranes
Enclose the material in the cell and regulate passage of material into the cell from the surrounding environment.

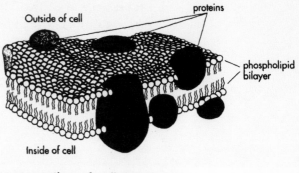

Cross-section of cell membrane
showing Fluid-mosaic nature

A. Fluid mosaic model

Describes a phospholipid bilayer as the foundation of a plasma membrane. The polar, hydrophilic end lies on the outside of the bilayer and the nonpolar, hydrophobic end lies on the inside. Embedded in the bilayer are cholesterol and a variety of protein molecules.

1. *Channel (transmembrane) proteins*—provide for the passage of certain water-soluble substances across the membrane.

2. *Electron transfer proteins*—transfer electrons from one molecule to another.

3. *Receptor proteins*—sites that act as receptors for trigger molecules that cause a cell response within the cell.

4. *Recognition proteins*—recognize and help in some cells sticking to other cells.

5. *Transport proteins*—use energy from ATP to transfer material across the membrane in a process called active transport.

B. Transport across the membrane

There are a variety of instances in which substances need to be transported across the membrane, both in and out. Some molecules diffuse through the membrane, and others require special transport proteins.

1. *Selectively permeable membrane*—any membrane through which only certain substances can pass. This process is dictated by several factors as explained below.

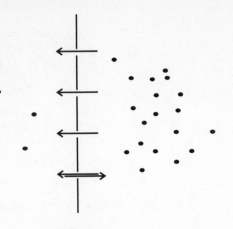

Diffusion

2. *Diffusion*—random movement of particles from an area of greater concentration to an area of lesser concentration, producing a diffusion gradient

3. *Osmosis*—diffusion of water

4. *Plasmolysis*—osmosis directed outward from the cell and leads to cell collapse

5. *Facilitated diffusion*—diffusion of water-soluble substances aided by channel proteins

6. *Active transport*—energy aided transport using ATP and transport proteins in the membrane

7. *Exocytosis*—movement of substances out of the cells within membrane vesicles, which release to the exterior of the cell

8. *Endocytosis*—movement of substances into the cell within the cell membrane

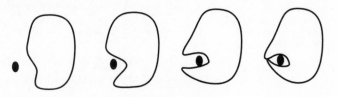

Phagocytosis

- Phagocytosis—plasma membrane engulfs an undissolved substance (usually whole cells or cell parts) too large to pass through the membrane.

- Pinocytosis—plasma membrane engulfs a dissolved substance—and therefore water as well—too large to pass through the membrane.

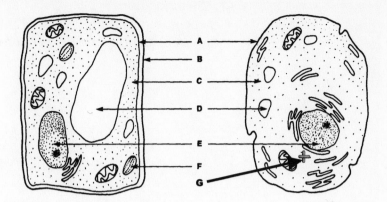

A - cell membrane
B - cell wall
C - cytoplasm
D - vacuole
E - nucleus
F - chloroplast
G - centrioles

Typical Cell Organelles

Cell organelles

Carry on the business of the cell through both membranous material and non-membranous material.

A. Cell walls

While not strictly an organelle, cell walls are found only in plants, fungi, some protists, and some bacteria; they consist mainly of cellulose in plants and help control osmotic uptake of water.

B. Centrioles

Give rise to spindle fibers. Almost exclusively found in animal cells.

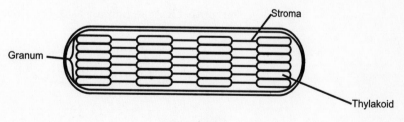

Chloroplast

C. Chloroplasts

The site for photosynthesis in plants cells, they contain photosynthetic pigments, including chlorophyll.

D. Chromosomes

Genetic material in the nucleus of the cell, they pass on hereditary information and regulate the cell through the production of proteins.

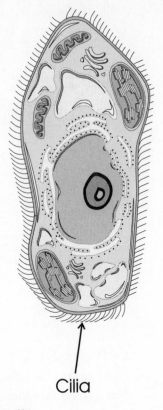

Cilia

E. Cilia and flagella

Have the same construction and aid in the movement of individual cells or movement of water past cells that are part of tissues, like lung epithelium.

F. Endoplasmic reticulum (ER)

Both smooth (no ribosomes embedded) and rough (ribosomes embedded) ER. The rough ER is the site where ribosomes make proteins for export from the cell with the aid of the various RNAs. Smooth ER makes lipids and detoxifies enzymes.

Flagella

G. Flagella

Longer than cilia but of the same material, they aid the cell in movement with a wave-like motion.

H. Golgi bodies

Also known as Golgi complex or apparatus, they are the site where proteins and other proteins are packaged for release from the cell.

I. Lysosomes

Contain digestive enzymes that metabolize substances for the cell, and even the cell itself, if it ages too much.

J. Microfilaments

Aid in the movement of substances in the cells as well as helping cells that change shape to move.

K. Microtubules

Provide support for the cell. Microtubules and microfilaments comprise the cytoskeleton of eukaryotic cells.

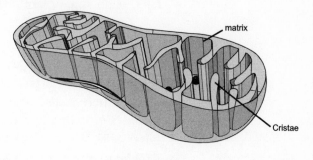

Mitochondrion

L. Mitochondria

The "powerhouse" of the cell, it is the sight of ATP synthesis only in aerobic respiration on the enfolded, inner membranes, which are known as cristae.

M. Nucleolus

A densely staining region of the nucleus where there are multiple copies of genes for making ribosomes.

N. Nucleus

A double-phospholipid-bilayer-bounded control center of the cell. It contains DNA, chromosomes that create proteins known as histones, and the nucleolus.

O. Peroxisomes

Organelles that break down substances in cells, usually toxins.

P. Ribosomes
The site of protein synthesis in the cell, they are composed of ribosomal RNA.

Q. Vacuoles
Usually found only in plant cells in the form of large storage areas.

Plant cells

Have an outermost cell wall that provides support but is primarily for osmotic control, chloroplasts where photosynthesis takes place, and a large vacuole filled with cell sap that contributes to the rigid nature of plant cells.

Animal cells

Have centrioles and take part in cell division; some have flagella or cilia, as do some plant cells, and some are able to engulf solid matter.

Cell cycles

There are three stages in the reproductive cycle of every eukaryotic cell: interphase, karyokinesis, and cytokinesis. In addition to these phases, there is also a sequence of events called the cell cycle that includes growth and replication of DNA. When cells divide to make other cells, it is necessary for them to copy their DNA so that daughter cells will each have a full set of information for controlling the cells' activities. The replication of eukaryotic cells is described by the cell cycle. In a typical multicellular eukaryotic organism, some cells will be actively dividing (in the cell cycle), such as skin cells in humans, and some will not. During the cell cycle, cells grow, replicate their DNA, grow some more, and then direct equal copies of their DNA (chromosomes) into two groups that will become the two nuclei of the daughter cells. Usually, the cell divides immediately (cytokinesis), but that process is not connected to mitosis. Often, mitosis occurs without cytokinesis, resulting in cells with more than one nucleus.

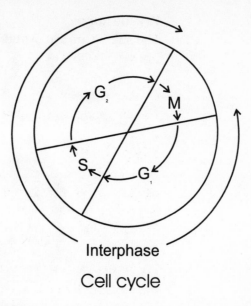

Interphase

Cell cycle

A. Interphase

Formerly thought to be the resting phase of mitosis since little could be seen happening, it is actually a very active phase in the life of the cell. It includes two growth phases, known as G1 and G2, and an S phase. Following these phases, the cell undergoes mitosis, or an M phase, when karyokinesis occurs, followed soon thereafter by cytokinesis.

1. *G1*—primarily a growth phase of the cell cycle.
2. *S*—growth and duplication of DNA.
3. *G2*—growth and preparation for karyokinesis.
4. *M*—the four stages of mitosis: karyokinesis followed by cytokinesis.

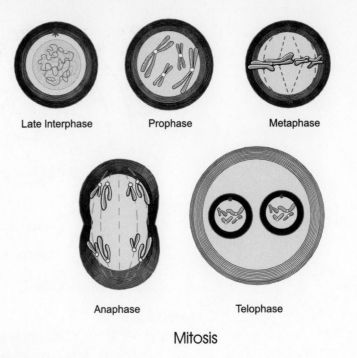

| Late Interphase | Prophase | Metaphase |

| Anaphase | Telophase |

Mitosis

B. Mitosis

Division of the nucleus of a cell that is identifiable in the life of the cell as four visible phases that take place: prophase, metaphase, anaphase, and telophase.

1. *Prophase*—activity in the cell is indicated by a disappearance of the nuclear membrane and the nucleolus, and condensation of the chromosomes begins and is completed by the end of prophase.

2. *Metaphase*—the chromosomes line up along the metaphase plate of the cell and become attached to the spindles at their central location, known as the centromere.

3. *Anaphase*—sister chromatids of each chromosome separate at their centromeres as the spindles begin to pull them to opposite poles of the cell.

4. *Telophase*—appears almost as the reverse of prophase. The chromosomes begin to unravel and become less visible in the two new daughter cells, as the nuclear membrane reappears around them. The nucleolus also reappears.

Peterson's AP Success: Biology

Cleavage furrow

Telophase in animal cell

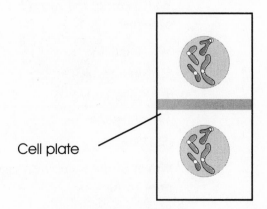

Cell plate

Telophase in plant cell

C. Cytokinesis

Following the stages of karyokinesis—during telophase specifically—in animal cells, the equator pinches at the cleavage furrow of the two new cells. In plant cells, a cell plate is formed between the two new daughter cells, ultimately producing a new cell wall between the new daughter cells.

MULTIPLE CHOICE QUESTIONS

1. Identify the correct eukaryotic cell cycle.

   (A) G1 to S to G2 to M to cytokinesis

   (B) G1 to G2 to M to S to mitosis

   (C) G1 to G2 to S to M to karyokinesis

   (D) S to G1 to G2 to M to cytokinesis

   (E) G2 to M to S to G1 to cytokinesis

2. Which of the following is NOT involved in the synthesis of proteins?

   (A) rough ER

   (B) smooth ER

   (C) Golgi body

   (D) ribosomes

   (E) RNA

3. Which of the following stages of the cell cycle is immediately prior to mitosis?

   (A) G1

   (B) G2

   (C) S

   (D) M

   (E) cytokinesis

Questions 4–6 refer to the following choices:

   (A) interphase
   (B) prophase
   (C) metaphase
   (D) anaphase
   (E) telophase

4. Chromosomes replicate.
5. Cytokinesis begins.
6. Chromosomes are pulled to opposite poles of the cell.
7. Which of the following produces ATP during aerobic respiration?

   (A) chloroplast

   (B) nucleus

   (C) ER

   (D) mitochondrion

   (E) ribosomes

8. If the concentration of a solute differs across a membrane permeable only to water, water will move across the membrane by

(A) facilitated transport

(B) osmosis

(C) phagocytosis

(D) active transport

(E) exocytosis

9. The _____ is a rigid structure that gives the cell support in plants.

(A) vacuole

(B) cell wall

(C) microfilaments

(D) nucleus

(E) centriole

10. Which of the following often distinguishes plant cells from animal cells?

(A) centrioles

(B) nucleus

(C) chromatin

(D) rough ER

(E) bilayer plasma membrane

EXPLANATION OF ANSWERS FOR MULTIPLE CHOICE QUESTIONS

1. **The correct answer is (A).** Following cytokinesis, the cell that continues in the cell cycle immediately goes through a growth phase (G1), followed by a phase that involves continued growth as well as replication of the DNA (S), followed by continued growth and preparation for mitosis (G2) after which mitosis—karyokinesis—occurs (M), often (but not always) ending with cytokinesis or dividing off of the cytoplasm between the two new daughter cells.

2. **The correct answer is (B).** Rough ER is the site for protein synthesis and has ribosomes embedded, which is why it appears "rough" in electron photomicrographs. The ribosomes, mainly composed of RNA, utilize other RNA—mRNA and tRNA—to synthesize proteins. Ribosomes on the RER produce proteins that are passed from the cell; the Golgi body then packages them for

this purpose. Smooth ER is used mainly in the making of lipids and detoxifying enzymes.

3. **The correct answer is (B).** Referring back to the answer for question number 1, it is noted that G2 is the phase just prior to the mitosis (M) phase. S is the phase where replication of DNA occurs. G1 phase occurs immediately *after* mitosis, and cytokinesis occurs at the end of mitosis, prior to G1, not mitosis.

4. **The correct answer is (A).** During interphase, which would include phases G1 to S to G2, the cell replicates its chromosomes. Specifically, it prepares for this in G1, it follows through with the replication in S, and it prepares for mitosis in G2. Choices (B) through (E) are the phases of mitosis. By this time, the chromosomes have to have already been replicated.

5. **The correct answer is (E).** Of all the phases of mitosis—choices (B) through (E)—telophase, which is at the end, is the end of karyokinesis and marks the imminent division of the cell or cytokinessis. Interphase is the growth and DNA replication phase. Chromosomes become visible during prophase, they line up along the equator of the cell in metaphase (Metaphase—Middle) and begin migrating toward the poles in anaphase.

6. **The correct answer is (D).** As is noted in answer number 5, the chromosomes are pulled, or migrate, to opposite poles during anaphase.

7. **The correct answer is (D).** The site for ATP production, of course, is on the inner membranes of a mitochondrion, known as the cristae. The ER houses the ribosomes and plays a role in protein synthesis, not ATP production. The nucleus is the site that houses the nucleic acids and is also where they are replicated. Chloroplast captures sunlight and traps it in high-energy carbon molecules, the energy of which is then released in the mitochondria to make ATP.

8. **The correct answer is (B).** The movement of water, by definition, across a selectively permeable membrane is by osmosis. Facilitated transport is the movement of substances—sometimes water and a solute—through channel proteins embedded in specific sites in the plasma membrane. Phagocytosis is the wholesale infolding of the plasma membrane to engulf particles too large to pass through the membrane. Active transport, of course, is the use of energy to facilitate the movement of a substance across the membrane, where the substance, without the addition of energy, would not normally be transported across the membrane. Finally, if the membrane is permeable only to water, the solute will not be transported across the membrane.

9. **The correct answer is (B).** The danger in picking vacuole for the answer is that it is the vacuole and the water in it that create turgor pressure, which does indeed provide for cell support in plant cells. However, taken alone—meaning specifically without water in it—the vacuole actually collapses (the typical wilting of some plants not watered for some time). Microfilaments don't provide for support; they aid in movement. Microtubules have a role in support as a cytoskeleton. RER and the nucleus have no rigidity to them. The cell wall is the constant source of rigidity in plant cells.

10. **The correct answer is (A).** Plant and animal cells both have the structures listed in choices (B) through (E). It is the almost total absence of the centriole in plant cells that distinguishes them from animal cells. The centriole plays a role in animal mitosis, and while plants lack this structure, they nonetheless also go through mitosis.

VOCABULARY

active transport
adenosine triphosphate
anaphase
carrier proteins
cell cycle
cell division
cell wall
cell sap
cell
centrioles
chloroplasts
chromatids
chromatin
chromosomes
cilia
cleavage furrow
cytokinesis
cytoplasm
cytoskeleton
diffusion
endocytosis
endoplasmic reticulum
eukaryotic
facilitated transport
flagella
fluid-mosaic model
G1 phase
G2 phase
Golgi bodies
homologues

interphase
karyokinesis
lysosomes
metaphase
microfilaments
microtubules
mitichondria
mitosis
nucleolus
nucleus
organelles
osmosis
passive transport
peroxisomes
phagocytosis
phospholipid bilayer
pinocytosis
plasma membrane
prokaryotic
prophase
ribosomes
S phase
sodium-potassium pump
spindle fibers
telophase
vacuoles
vesicles

Chapter 3

ENERGY FOR LIFE—ANABOLISM AND CATABOLISM

OVERVIEW

The currency of the cell is ATP, and a study in bioenergetics must involve a substantial knowledge of how this molecule—and by default ADP—works. The student should study this chapter to achieve an understanding of the importance of coupled reactions: energy released from one is used to drive another reaction. ATP as a source of activation energy in many reactions is a good example. Oxidation as a source of energy to pump H+ ions across a membrane is another. This section briefly describes the formation and breakdown of biological substances and then launches into discussions of, first, anabolism and then catabolism. The AP student, of course, will not fall into the trap that portrays autotrophs as carrying on photosynthesis only and heterotrophs as carrying on respiration only. Autotroph translates roughly into "makes own food." Use of the word "food" indicates that they use the food they make as well; heterotrophs are just fortunate enough to be able to use it also. Autotrophs carry on photosynthesis and respiration—they still need energy when the sun goes down—and heterotrophs carry on respiration only. If there is a group of organisms dependent on another, it is the heterotrophs. To begin our discussion, we start with the building of energy-rich (energy-stored) molecules.

Starch polymerization (synthesis)

ANABOLISM

Synthesis—or building—of molecules to higher-ordered energy states—enthalpy—leading to greater free energy, or ΔG, and resulting in more chemical organization. Photosynthesis is one of the primary anabolic series of reactions that result in energy-rich molecules assembled to yield glucose. This occurs in many autotrophs, paving the way for formation of all of the organic molecules of the organism. Any reaction that builds a substance with higher chemical organization than its component parts is anabolic, such as the formation of macromolecules in the production of cell parts and in the production of larger biological molecules from elements such as C, H, N, O, P, S, etc. Many large biological molecules are the result of the breakdown of even larger molecules, so the student should distinguish between the molecule and the process.

$$C_{20}H_{40} + 60O_2 \longrightarrow 20CO_2 + 20H_2O + energy$$

Catabolism

CATABOLISM

Breakdown of large molecules to smaller molecules, lower-ordered energy states—entropy—leading to less free energy, which results in less chemical organization. Respiration, both aerobic and anaerobic, is one of the primary series of catabolic reactions that result in the release of raw materials that can again be used to supply anabolism and the release of free energy to do work. One example would be the storing of free energy in ATP. The reactions of respiration—catabolic in nature—are coupled with the production of ATPs in the system—anabolic in nature.

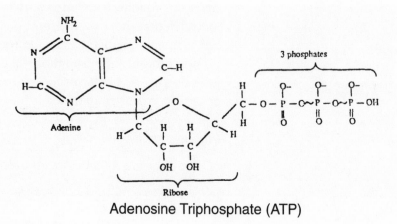

Adenosine Triphosphate (ATP)

ATP

Provides the energy for many biological chemical reactions. It is composed of three distinct parts: a 5-carbon sugar called ribose, an adenine base, and three phosphate groups aligned in a linear fashion, extending from the central, ribose sugar.

PHOSPHORYLATION

Outer two phosphates are linked to the molecule with high-energy bonds. When the outermost bond breaks, energy is released for some cell processes and ADP forms with a phosphate group, usually bound to an intermediate substrate in a process called phosphorylation.

SOURCES

It is the formation of these bonds that occurs in the catabolism of energy-rich carbon molecules during respiration. The ATP that is formed in photosynthesis is primarily a carrier molecule for energy.

PHOTOSYNTHESIS

Production of high-energy molecules in the presence of light, using carbon dioxide, water, chlorophylls, ATP, NADP, and enzymes. A change in any of these ingredients will, of course, change the rate of the reaction. The overall reaction is given by

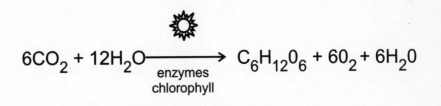

$$6CO_2 + 12H_2O \xrightarrow[\substack{\text{enzymes} \\ \text{chlorophyll}}]{} C_6H_{12}O_6 + 6O_2 + 6H_2O$$

Summary of Photosynthesis

CHLOROPHYLL

Is actually several light-absorbing pigments—chlorophylls a and b—found in chloroplasts that are involved in absorbing light energy, which is then passed on to the reaction centers of the photosystems through the use of antenna chlorophylls.

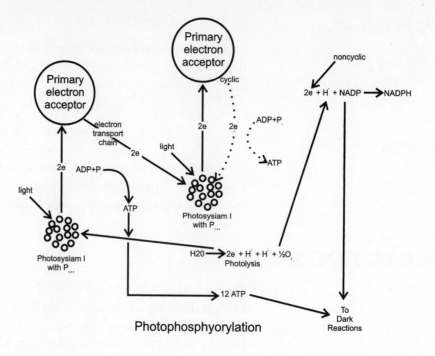

Photophosphyorylation

LIGHT REACTIONS

Comprised of reactions that occur in the presence of light in what are called photosystem I and photosystem II. These reactions carry on the phosphorylation that produces ATP. The LR take place in the thylakoids located in the grana of the chloroplasts.

Photosystems I (PS I)

Reaction center contains a chlorophyll molecule, known as P700, which absorbs light energy in the 700-nanometer range and reactivates energy-drained electrons from PS II. These activated electrons reduce $NADP^+$ to NADPH, a carrier molecule that transports both energy and hydrogen to the carbon fixation reactions in what are sometimes known as the dark reactions. It takes part in both cyclic and noncyclic photophosphorylation.

Photosystems II (PS II)

In this reaction center, the chlorophyll molecule known as P680, which has an absorption peak in the 680-nanometer range, is a system by which electrons are activated to help form ATP in the process known as photophosphorylation. The P680 center also takes part in photolysis, which splits water into hydrogen and oxygen ions and replaces the missing electrons from PS I.

Photophosphorylation

The addition of a phosphate to an ADP molecule making it an ATP molecule. ATP then transports this energy from the light reactions to the dark reactions—the fixation of carbon. As the name suggests, it takes place in the presence of light.

A. Noncyclic

Drives an activated electron to the dark reactions using the energy-carrying molecule ATP and the energy and hydrogen-carrying molecule of NADP, utilizing both the PS II and the PS I reactions. The electron is not recycled but is replaced by another electron from the ionization of water in PS I photolysis, which produces the hydrogen ions used to make $NADPH_2$ and releases free-oxygen molecules. Thus the electrons are passed from the light reactions to the dark reactions using $NADPH_2$.

B. Cyclic

Involves only the PS I reactions, wherein the electrons leave the reaction center and are passed through a system of carrier molecules and eventually are returned to PS I. ATP is produced at the very end of this cycle. Plants use this relatively inefficient series of reactions when they lack sufficient NADP.

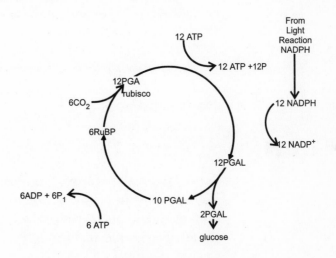

Calvin-Benson cycle

Dark Reactions

So-called because they do not utilize the energy directly from light but rather use the energy-rich molecules formed during the light reactions. Atmospheric carbon dioxide is the source of carbon for the ultimate formation of the energy-rich carbohydrate molecules at the end of the dark reactions known as carbon-fixation.

Calvin Benson (C_3) cycle

Carbon dioxide enters the Calvin-Benson cycle by combining with a 5-carbon sugar known as ribulose biphosphate (RuBP, see above). RuBP and the "fixed" CO_2 form an unstable compound and thus the cycle continues with the splitting of RuBP into two 3-carbon—thus the C_3 pathway—molecules known as phosphoglyceric acid (PGA), which converts eventually to phosphoglyceraldeyde (PGAL) that is ultimately converted to the 6-carbon molecule glucose.

C_4 pathway

In hot, dry climates, plants have evolved a very efficient method of sugar production that conserves water called the C_4 pathway, since it initially involves a 4-carbon molecule. The C_4 pathway occurs in the stroma of the chloroplasts.

CAM pathway

Another efficient method of fixing carbon in some climates with hot, dry days and cool nights, this is almost identical to the C_4 pathway except that malic acid is shuttled to the vacuole and the stomata are open at night and closed during the day, thus conserving water.

$$C_6H_{12}O_6 + 6O_2 \xrightarrow[\text{cytochromes}]{\text{enzymes}} 6CO_2 + 6H_2O + ATP$$

Summary of Aerobic respiration

RESPIRATION

Catabolic (breakdown) reaction that releases the stored energy in glucose so that it might be stored in ATP. In many ways cellular respiration and photosynthesis are parallel yet are opposite reactions. Cellular respiration starts with the high-energy, multicarbon molecule glucose and breaks it down to carbon dioxide, water, and ATP using oxygen, some ATPs to prime the pump, and enzymes. It, too, uses electron transport chains and ends with a substantial production of ATPs and water.

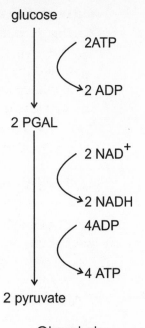

Glycolysis

GLYCOLYSIS

After passing across the plasma membrane with the help of several proteins in the membrane, glucose is activated in the cytoplasm and transformed into a sugar diphosphate, which degrades to two PGAL molecules that are the next step in the formation of pyruvic acid. This is the beginning of anaerobic respiration since no oxygen is invested and results in a net gain of two ATPs—two ATPs start glycolysis and four are produced. If oxygen is available at this time, the pyruvic acid enters the mitochondrion, where it continues degradation until the final result is the bulk of the ATP production, water, and $FADH_2$. If oxygen is not available, aerobic respiration soon comes to a halt, with the result being a multicarbon molecule still rich in energy.

AEROBIC PROCESSES

Continued breakdown of glucose to completion, fueled by the presence of oxygen. This phase is carried out on the enfolded membranes of the mitochondrion called cristae. The net result is a large amount of energy in the form of ATP for the cell to carry on its work.

Acetyl-CoA

Upon entering the mitochondrion, the 3-carbon pyruvic acid is converted to Acetyl Coenzyme A by forming a substrate complex with Coenzyme A. Essentially, this delivers the degraded pyruvate to the Kreb's—citric acid—cycle, where it undergoes a series of reactions that result in the production of some ATP, FADH$_2$, and NADH.

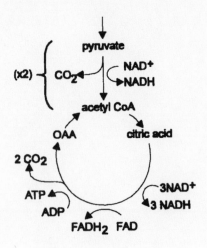

Kreb's (citric acid) Cycle

Kreb's (citric acid) cycle

Coupling the end product of the last turn of the Kreb's cycle, a 4-carbon molecule known as oxaloacetate, the 2-carbon Acetyl-CoA enters the Kreb's cycle. A chain of reactions results in the formation and then degradation of various multicarbon molecules (see below), the formation of NADH and FADH$_2$, and the release of two carbons as carbon dioxide. The end result of this turn of the cycles is the above-mentioned substances and oxaloacetate, which is precisely where the cycle began. The next Acetyl-CoA couples with this oxaloacetate and the cycle turns again. The resulting NADH and FADH$_2$, still rich in energy and hydrogen, now enter the electron transport chain, with the result being the bulk of the ATP production.

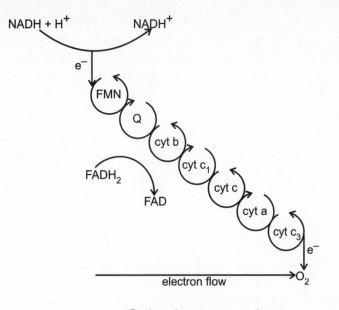

Cytochrome system

Electron Transport

Hydrogen and its energy are transferred to carrier molecules. They are then shuttled to the inner membranes of the mitochondrion, where the electrons undergo a series of reactions. Traveling down the electron transport chain, their energy is coupled with the final electron acceptor, oxygen, which, along with the hydrogen from oxidative phosphorylation, forms the bulk of the water from cellular respiration.

Oxidative Phosphorylation

Energy from the electron transport chain is used to pump the hydrogen ions across the inner membranes of the mitochondrion creating a hydrogen (proton) gradient. This causes the hydrogen (proton) to eventually cross back over the membrane, resulting in a large production of ATP and providing the hydrogen ions needed to form water with the activated oxygen ions from the electron transport chain

ANAEROBIC PROCESSES

Results when no oxygen is available, either by an accident of nature or by design, as in brewing of alcoholic beverages. When cells enter anaerobic respiration, the result is a molecule still very high in energy and some carbon dioxide, depending on the fermentation schema. This takes place entirely in the cytoplasm, and the cell releases alcohol in plant tissue and lactic acid in mammalian tissue, both generally considered toxins for the cells.

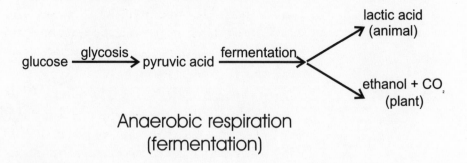

Anaerobic respiration (fermentation)

Plant

Results in the production of alcohol in plant tissues and the release of carbon dioxide.

Animal

Results in the production of lactic acid, for example, in human tissue.

Enyzyme + Substrate ⟶ Enzyme Substrate + Reactant Complex ⟶ Enzyme + Product

Enzyme-Substrate action

ENZYMES

Catalysts, both inorganic and organic, change the rate of a chemical reaction, most often speeding it up in a biological system, without changing as a result of the reaction. Instead, they are seen as organic facilitators. Enzymes are protein molecules that do just such a thing in biological systems and are, therefore, seen as organic catalysts. There are very specific, work with other chemicals that they bring changes in, are affected by a variety of environmental factors, and often work with a coenzyme.

SPECIFICITY

Enzymes are very specific to the chemical that they are helping undergo a change. Enzymes act on only one chemical.

SUBSTRATES

Enzymes work with substrate(s) to form a needed product. The substrate(s) can be seen as the reactants in the reaction that, as a result of the enzyme-substrate complex formed, most often have their activation lowered so the reaction will more easily and rapidly take place. The end result is the enzyme and the product. The enzyme is now free to carry out the reaction again, unless an inhibiting molecule attaches to it and prevents it from doing so (see Chapter 1).

COENZYMES

Couple with the enzyme to aid in the job of the enzyme. Vitamins are often coenzymes in some reaction vital to the system. Inorganic substances, such as several metallic ions, also serve as coenzymes.

FACTORS AFFECTING

Enzymes are affected by several environmental factors that speed up its action, slow it down, or stop it altogether. Some of the factors, in the extremes, effectively cause irreversible reactions. Fried eggs cannot be returned to the liquid state.

Temperature

Since temperature increases the motion of particles leading to the increased potential of collisions, this factor will increase the reaction as the temperature increases and slow it down as the temperature decreases. Refrigeration helps slow down the enzymatic reactions in microbes that can lead to food spoilage.

pH

Most enzymes work best around a neutral pH of 7. An exception would be the enzymes in the stomach, which, aided by gastric juices, work best around a pH of 2 to 3.

Substrate concentration

Amount of available substrate controls the rate of an enzyme-driven reaction. As the amount of substrate increases, the number of enzyme-substrate collisions increases, driving the rate of the reaction to occur faster. If enough substrate binds to the active site of an enzyme, the reaction will plateau—reach equilibrium.

MULTIPLE CHOICE QUESTIONS

1. Which of the following is true about an enzyme?

    (A) They always work alone.

    (B) They are consumed in a reaction.

    (C) They are amino acid polymers.

    (D) They always require a coenzyme.

    (E) They are classified as inorganic catalysts.

2. Which of the following has a vitamin as a building block?

    (A) apoenzyme

    (B) alloenzyme

    (C) metallic ion

    (D) lipoprotein

    (E) coenzyme

3. Which enzyme would the microbes in the gut of a termite need to have in order to metabolize the cell walls of the wood that termites eat?

    (A) cellulase

    (B) esterase

    (C) protease

    (D) pepsin

    (E) trypsin

4. The reds, oranges, and yellows of the leaves of deciduous trees that become evident in the fall are from

    (A) carotenoids

    (B) ATP

    (C) leaf decay

    (D) chlorophylls

    (E) overabundance of water

5. Which of the following is a product of cyclic photophosphorylation?

    (A) carbon dioxide

    (B) oxygen

    (C) ATP

    (D) NAD^+

    (E) Acetyl CoA

6. The role of oxygen in aerobic respiration is

   (A) to couple with C to form CO_2 in chemiosmosis

   (B) to form ATP

   (C) to contribute H^+ to the Kreb's cycle

   (D) used to make PGAL

   (E) used to accept electrons from the ETC

7. Photosynthesis is

   (A) light

   (B) driven

   (C) anabolic

   (D) enzyme moderated

   (E) A and C only

   (E) A, B, and C

Questions 8 through 10 use the following choices:
   (A) stroma
   (B) thylakoid
   (C) vacuole
   (D) stomates
   (E) lamella

8. The light reactions of photosynthesis take place here.

9. The Calvin cycle of photosynthesis take place here.

10. The site of carbon fixation.

EXPLANATION OF ANSWERS FOR MULTIPLE CHOICE QUESTIONS

1. **The correct answer is (C).** If choice (A) excluded the word "always," it might be an attractive choice. Enzymes, as has been noted, are not changed by the reaction they facilitate, which excludes choice (B). Very often, they do require a coenzyme, but, once again, the word "always" excludes some enzyme-mitigated reactions. It is very definite that the definition of enzymes calls them organic catalysts. Inorganic catalysts are an entirely different class of catalysts with different physical properties, although they still affect the rate of a reaction.

2. **The correct answer is (E).** All enzymes are composed of protein that form one of two binding sites. One is for the allosteric binder and one is for the substrate, and this enzyme is known as an allosteric enzyme. Also, the more complex the

enzyme, the more cofactors—nonprotein parts—it has. If the cofactor is an easily removable cofactor, it is a coenzyme. Vitamins qualify for this designation, but metallic ions do not since they bond quite securely. If the enzyme is without the coenzyme, it ceases to function. Apoenzymes do not exist.

3. **The correct answer is (A).** The enzyme in the gut of a termite has to metabolize the cell wall of the wood cells, and those walls are composed of cellulose. Enzymes that break down substances have the suffix *-ase* attached to them, so the enzyme for metabolizing cellulose would be cellulase. Proteases metabolize proteins; pepsin and trypsin are proteases.

4. **The correct answer is (A).** Chlorophylls are active during the growing season and are eventually dismantled at the end of that period in the life of deciduous trees. This paves the way for the red, orange, and yellow carotenoids to be seen. Leaf decay, as in perhaps a dead leaf, would take on a brown appearance, and ATP is certainly not responsible for any colors in a plant. An overabundance of water might lead to bursting of the cells, but this would not be accompanied by the wholesale color changes we see in deciduous trees in the fall. It is important to note that the carotenoids are not produced in the fall; they are there for the life of the leaf.

5. **The correct answer is (C).** Cyclic photophosphorylation occurs only when there are not enough NADPH molecules around to accept electrons. It is fairly inefficient and doesn't occur often; the abundance of water would obviously have an effect on this process. With little or no NADP available, NAD^+ is not produced or even used in cyclic photophosphorylation. Acetyl-CoA is involved in the Kreb's cycle of cellular respiration where no phosphorylation due to light occurs. The release of oxygen occurs in the photolysis of water during the several reactions that make up non-cyclic photophosphorylation. Carbon dioxide is not produced during photosynthesis; it is consumed to produce carbohydrates. ATP is the only thing produced in this reaction.

6. **The correct answer is (D).** The primary role of free oxygen in respiration is to accept electrons at the end of the ETC following, which it couples with H^+ to produce water at the end of aerobic respiration. Chemiosmosis involves the pumping of hydrogens across the thylakoid membranes in the production of ATP in photosynthesis. Oxygen does not play a role in contributing H^+, nor does it contribute to the synthesis of PGAL, the molecule in the Calvin-Benson cycle of photosynthesis.

7. **The correct answer is (E).** Photosynthesis is all of the choices (A) through (D). It is first and foremost driven by the photons that are the energy packets given off by a source of light. It is anabolic since it produces a higher energy state, which is composed of energy-rich molecules that represent more ordered systems of energy and, therefore, have higher free energy. Enzymes are, of course, crucial to this process and are, as a whole, non-cyclic in nature, proceeding from CO_2 and H_2O to $C_6H_{12}O_6$, a sequence that does not cycle. There are cycles within photosynthesis, but the reaction as a whole is non-cyclic.

8. **The correct answer is (B).** The oxidation-reduction reactions involved in oxidation of water, reduction of $NADP^+$, and synthesis of ATP via a chemiosmotic mechanism are located in the thylakoid membranes of the chloroplast. The soluble enzymes that are involved in the synthesis of glucose from carbon dioxide and water and utilize the ATP and NADPH produced during the light reactions are in the stroma compartment of the chloroplast.

9. **The correct answer is (A).** When ATP and $NADPH_2$ are produced in the thylakoids at the end of the light reactions, they proceed to the stroma, where they drive the C_3 cycle.

10. **The correct answer is (A).** Carbon fixation occurs during the Calvin-Benson cycle of the dark reactions, which means this also occurs in the stroma.

VOCABULARY

acetyl coenzyme A
activation energy
active site
aerobic respiration
allosteric site
anaerobic respiration
antenna pigments
ATP
ATP-synthase
autotrophs
bioenergetics
C4 pathway
Calvin cycle
carbon fixation (C3 pathway)
carotenoids
cellular respiration
chlorophyll a
chlorophyll b
chloroplasts
citric acid cycle
coenzyme
cofactor
competitive inhibition
conducting tissue
cuticle
cyclic photophosphorylation
cytochromes
electron transport chain
entropy
enzyme specificity
enzyme-substrate complex
enzymes
ethanol
feedback inhibition
fermentation
glycolysis
glyceraldehyde-3-phosphate (G3P)
grana
guard cell
heterotrophs
hydrogen ion gradient

induced fit
Kreb's cycle
lactic acid
light reaction
lock-and-key theory
lower epidermis
metabolic cascade
mitochondria
NADPH
noncompetitive inhibition
noncyclic photophosphorylation
nonspontaneous reaction
organic catalyst
oxaloacetate
oxidative phosphorylation
p680
p700
palisade cells
phosphoglyceric acid (PGA)
phosphophenolpyruvate (PEP)
photolysis
photon
photophosphorylation
photosynthesis
photosystem I
photosystem II
pyruvic acid
reaction center
ribulose biphosphate (RuBP)
spongy mesophyll cells
spontaneous reaction
stomate
stroma
substrate
thermodynamics—first law
thermodynamics—second law
thylakoids
upper epidermis

Unit II—Heredity and Evolution

Chapter 4

GENETICS: MENDEL AND HEREDITY

OVERVIEW

Since Mendel's time, scientists have added to his fundamentals and have linked such processes as meiosis as well as mitosis to his findings. Modern genetics has explored crossing over, nondisjunction, incomplete dominance, linkage, and the like. Our current knowledge of these areas has even engendered the field of genetic counseling. Use of the Punnett square is assessed on the exam, and the student should know how to utilize this tool in calculating answers in genetics, particularly Mendelian genetics. The student should know these processes in detail as well as Mendel's laws, be capable of comparing and contrasting mitosis and meiosis (a favorite essay question), know the differences between spermatogenesis and oogenesis, and be able to perform monohybrid crosses, test crosses, and dihybrid crosses. The student should also be comfortable with the basics of Mendel's work and be able to transition to modern genetics—the subject of Chapter 5—with confidence. Finally, the study of the patterns of inheritance will be discussed from a variety of levels and is a theme that actually runs through most of the study of this area.

MENDEL

The "father" of genetics and a nineteenth-century monk trained in mathematics, Mendel conducted years of studies on the traits of pea plants as a hobby. He theorized that certain "factors"—known today as genes—controlled the passing on of these traits from one generation to the next. His laws are the foundation of genetics and are the principles upon which we base our current knowledge of inheritance. His techniques were precise, and his knowledge of mathematics was an essential asset to his conclusions. In discussing terminology, the student should be aware that not all terms listed emanated from Mendel's original work. A large measure of them are "post-Mendelian" but involve either clarification of his work or the discovery of new principles that dictated new terms. We begin with a discussion of his laws; meiosis is treated separately but is obviously an integral part of these laws and, in fact, embody what Mendel discovered.

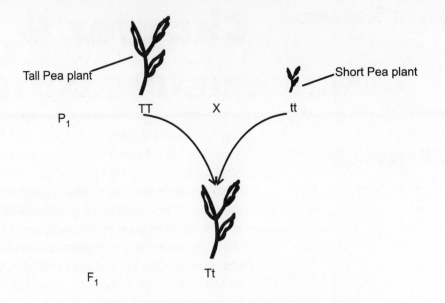

Mendel's Law of Dominance

LAW OF DOMINANCE

First and foremost, Mendel noticed that certain traits did not have a halfway point or a blending affect (see Incomplete Dominance). Such was the case with the trait for height in peas in a monohybrid cross or mating for one trait. Mendel's conclusion was that in the expression of the trait for height in peas, tall pea plants dominate over short pea plants; Mendel called them factors, and we now call them genes or alleles. When a pea plant homozygous, or pure, for tallness is crossed with a pea plant homozygous for shortness, it appears that the offspring produced in the F_1, or first filial generation, have lost the recessive allele. This is an indication that one trait is dominant, and the other (seemingly vanished) trait is recessive. Eventually, Mendel came to conclude that the recessive trait was being masked. In truth we know today that, although each offspring receives one gene from each parent (the P_1 generation) for a trait, only one gene works to express that trait. In the case of a homozygous individual, it doesn't matter which one because they are identical. In a heterozygous individual, it will be the dominant allele. Thus, the effect of the Law of Dominance was a powerful one, illuminating the relationship between dominant and recessive alleles. It should be noted that farmers and agriculturists were aware of this relationship to a certain extent before Mendel but used it informally in a hit-or-miss manner. Mendel codified the relationship, and, although his work sat on a library shelf for more than three decades, it nonetheless had a powerful effect on the world when it was rediscovered.

LAW OF SEGREGATION

Prior to the potential for mating, certain cells go through a process we now know as meiosis. During this time, the alleles ultimately separate from each other and end up in separate gametes or sex cells. Mendel saw this as a mathematical construct and realized that anything else would be mathematically impossible. If all the genes of an individual were passed on each time mating occurred, the number of genes each successive generation receives would double. If instead, as we now know occurs in meiosis, the number of alleles is halved through segregation from each other before they are "doubled" at mating, the number of alleles in each generation remains constant. In using the Punnett square, we put one allele above each column or on each row to signify that the genes must separate from each other before they can reunite in sexual reproduction.

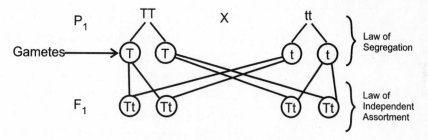

Mendel's Law of Segregation and
Law of Independent Assortment

LAW OF INDEPENDENT ASSORTMENT

Simply stated, this law says that the way that alleles for a trait are "coupled" to each other in sexual reproduction is random and all recombinations are possible, with every mating. For example, when we cross two F_1 offspring from the mating of a homozygous tall and a homozygous short. (NOTE: When referring to the recessive trait, it is not necessary to use the term homozygous: a recessively expressed phenotype (physical appearance as a result of possessing certain alleles) means that the individual *has* to be homozygous.) The presence of a dominant allele means that the individual will not express the recessive trait due to the presence of that dominant trait. A favorite ruse of the test-maker is to appear to be giving less information about the genotype (the exact nature of the alleles, TT, Tt or tt) than is needed when referring to the individual whose phenotype expresses the recessive trait. Indicating that the individual's phenotype is short, as in the case of peas, without indicating homozygous or heterozygous short does this. In order for a recessive trait to appear, it must be homozygous, thus the term homozygous is irrelevant when referring to the recessive phenotype, and the use of the term heterozygous would be incorrect. In the case of peas, a phenotype of tall, on the other hand, must be accompanied by a description of the genotype—either homozygous {TT} or heterozygous {Tt}. In that cross, the possible offspring could be {TT}, {Tt}, {Tt}, or {tt}. A better illustration of independent assortment is found in a dihybrid cross where, for example, if the alleles are A, a, B, and b, the possible matings are {AB}, {Ab}, {aB}, or {ab}.

	T	t
T	TT	Tt
t	Tt	tt

Monohybrid Cross

PUNNETT SQUARE

Used to better organize the crosses and their possible outcomes between two individuals. The use of capital letters in the illustrations represents the law of dominance. Placing each allele in a separate column or row represents the law of segregation. The inner squares of the Punnett square represent the law of independent assortment, more aptly called the law of independent "resortment." All three of Mendel's laws can be illustrated in a Punnett square.

MONOHYBRID CROSSES

Mating of individuals that are the P_1 hybrid for one trait. This is originally a reference to the dominant/recessive state between individuals within a species and does not hold for incomplete dominance, although the laws of Mendel hold true for both. Several monohybrid crosses are indicated below, including a test cross. (NOTE: the only way a test cross can successfully reveal the presence of a recessive gene in a dominantly expressing individual is to cross that individual with a recessively expressing individual. If there is a recessive gene being masked in the dominant individual, it will only be expressed if that individual is mated with a recessive individual for that trait. Three test crosses are illustrated below.)

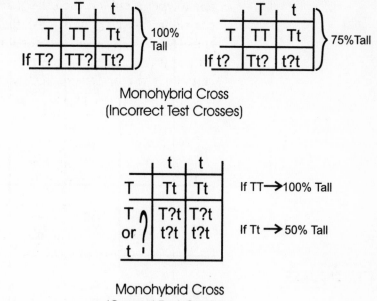

Monohybrid Cross
(Incorrect Test Crosses)

Monohybrid Cross
(Correct Test Crosses)

Test Crosses in Monohybrids

Monohybrid does not refer to the fact that each member involved in the cross need be hybrid. It refers to the fact that from individuals homozygous for both the dominant and the recessive trait, when crossed, will result in hybrids. Another way to say it is that monohybrid crosses reflect one of each trait, the dominant and the recessive. We also use this to refer to a hybrid individual who has one dominant allele and one recessive allele. The classic monohybrid results of a phenotypic ratio of 3:1 and genotypic ratio of 1:2:1 are noted below.

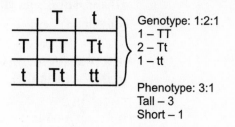

Genotype: 1:2:1
1 – TT
2 – Tt
1 – tt

Phenotype: 3:1
Tall – 3
Short – 1

Monohybrid Cross Results

	AB	Ab	aB	ab	Genotype
AB	AA BB	AA Bb	Aa BB	Aa Bb	AABB: 1 AABb: 2 AaBB: 2 AaBb: 4
Ab	AA Bb	AA bb	Aa Bb	Aa bb	AAbb: 1 Aabb: 2 aaBB: 1 aaBb: 2 aabb: 1
aB	Aa BB	Aa Bb	aa BB	aa Bb	Phenotype A+B trait: 9 A+b trait: 3
ab	Aa Bb	Aa bb	aa Bb	aa bb	a+B trait: 3 a+b trait: 1

Dihybrid Cross

DIHYBRID CROSSES

A cross potentially hybrid for two traits, illustrated below, this type of a cross still holds true to Mendel's laws, there are simply more gametes possible and more offspring. The typical true genotypic ratio is cumbersome to note here, although it is illustrated, and the typical true dihybrid phenotype ratio is the classic 9:3:3:1. If the traits are *A* or *a* and *B* or *b*, the phenotypic nine would produce all *AB* appearing, the first three might be *Ab* appearing, making the second three *aB* appearing, and the one would be represented by *ab* appearing individuals.

```
RR   X   WW          RW   X   RW
(red)   (white)      (pink)   (pink)
       |                     |
       v                     v
      RW            RR      RW      WW
    (pink)        (red)   (pink)  (white)
     100%          25%     50%     25%
```

Incomplete Dominance

INCOMPLETE DOMINANCE

Characteristics intermediate between parental extremes are shown; the traits combine to produce an intermediate form indicating that neither is dominant, thus the name. Instead of making up a new symbol to indicate the third trait, it is illustrated by showing it as being made up of one each of the two alleles in question as illustrated. Care should be taken on questions referring to incomplete dominance. The greater number of "blended" offspring would be given not by a cross between two of these individuals but by crossing individuals pure for the component alleles. If *R* represents red in four-o'clock flowers, *W* represents white, and *RW* represents pink, the largest number of pink flowers would come from mating a red [*RR*] and a white [*WW*] four-o'clock flower plant. By Mendel's laws, all of the offspring would be *RW*.

MEIOSIS

Stages in the life of cells as they prepare to pass on their traits to the next generation. Meiosis has two divisions in contrast to the one division of mitosis. Meiosis is often referred to as a reduction division. Meiosis is the embodiment of Mendel's Law of Segregation. As will be seen, this is where crossing over, nondisjunction, and the like do matter. A chromosome is composed of two identical "sister" chromatids. The are identical to each other and are in fact duplicates. Only one half is present in the normal workings of the cells. An exact copy is made and attached to its "sister" half during interphase in order to prepare for the division about to occur.

CELL TYPES

When referring to cells and their division in meiosis, we use two terms, depending on how many of the necessary compliment of chromosomes each cell has. One refers to the number we typically find in body, or somatic, cells, and the other is used to refer to the number of chromosomes found in the sex cells located in the primary sex organs, or gonads, of the female (the ovaries) or the male (the testes).

Haploid
Refers to half of the number of the chromosomes normally found in the body cells of an individual. This number is achieved through meiosis and occurs mainly before all cells, with the exception of plants, that are ready for reproduction. Sexual reproduction brings the cell back to the diploid number.

Diploid cells
Refers to the normal chromosomal number found in the body cells of individuals capable of sexual reproduction. Also found in the sex cells prior to meiosis.

STAGES

Meiosis goes through two divisions, with the stages named after similar stages found in mitosis but labeled with a I or II. Condensation of the chromosomes occurs and they migrate to the middle of the cell, are separated, and migrate to opposite ends of the cell. This is where the similarity ends, however, between mitosis and meiosis.

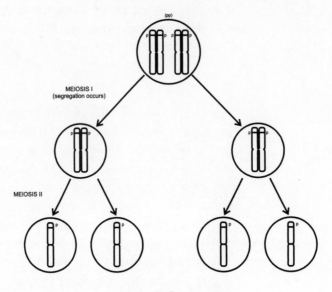

Meiosis Summary

Meiosis I

Following the disappearance of the nuclear envelope, chromosomes begin to condense, and homologous pairs lie next to each other in a manner similar to mitosis. The way in which they line up and what can happen to them, however, is very different in meiosis I.

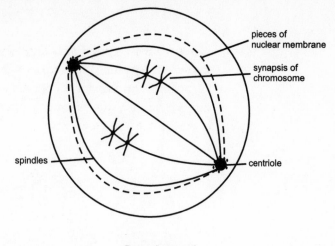

Prophase I

A. Prophase I

Homologous pairs begin to line up next to each other as pairs—not as individuals as in mitosis—in a process called synapsis. This makes four entities, which explains why they are called tetrads at this point. Homologous pairs means that they are biochemically very similar: the genetic code for blue eyes versus brown eyes starts in exactly the same way, although the code for those colors is chemically different beyond those starting points. If these closely associated segments by virtue of their chemistry happen to intertwine in close proximity, they may exchange parts in a process called crossing over. Occasionally, homologous pairs do not separate, or they separate inconsistently in a process know as nondisjunction. This will most likely cause more information to be passed on in one or more of the resulting sex cells.

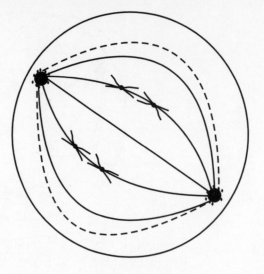

Metaphase I

B. Metaphase I

Tetrads line up along the equator in the middle of the cell as in mitotic metaphase. Once again, they are lined up as pairs, not as individual chromosomes as in mitosis. Spindles are attached to the centromeres and the homologous pairs, with their sister chromatids attached, and are about to be separated, providing the basis for the variation noted in the Law of Independent Assortment.

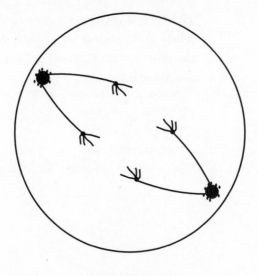

Anaphase I

C. Anaphase I

The homologous pairs of each tetrad begin migrating to opposite ends of the cell, similar to mitosis. At this point, the number of chromosomes in each forming cell is reduced by half.

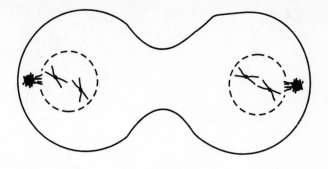

Telophase I

D. Telophase I
The nuclear envelope reforms around each set of chromosomes and the cells divide (cytokinesis).

Meiosis II
This part of meiosis, using only one half of the tetrad, though, is almost identical to mitosis.

A. Prophase II
Chromosomes condense and become visible as sister chromatids attached at the centromere but only as half the number of normal homologous pairs.

B. Metaphase II
The chromosomes line up along the equator of the cell.

C. Anaphase II
The chromosomes split at the centromeres and begin migrating to opposite ends of the cell.

D. Telophase II
The nuclear envelope reforms, and the result is four haploid cells.

GAMETOGENESIS

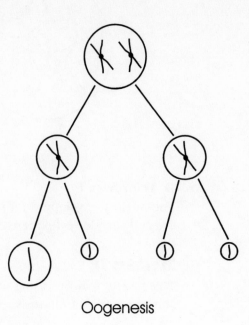

Oogenesis

Oogenesis

In females, gametogenesis produces eggs, or ova, and is called oogenesis. For every diploid cell that undergoes gametogenesis in females, one egg is produced. In oogenesis, the other three cells receive very little cytoplasm and eventually degenerate in a move that conserves cytoplasm for the surviving cell.

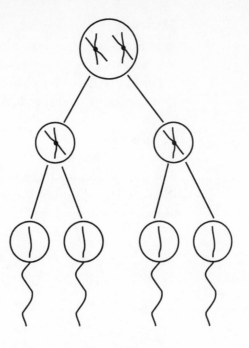

Spermatogenesis

Spermatogenesis

In males, gametogenesis produces sperm and is called spermatogenesis. For every diploid cell that undergoes gametogenesis in males, four sperm are produced.

CHROMOSOME INTERACTIONS

During the course of both mitosis and meiosis, the chromosomes come in close proximity and, since they are composed of very similar chemical material, they may interact. The bulk of this section is spent on the matter at hand, namely meiosis.

Mitosis

In body cells, there is no risk of what the next generation of individuals will receive as genetic information. Any interaction of chromosomes in mitosis, other than what usually should happen, will lead the cells to die off or pass on the anomaly, called a mutation, which is usually negative. (NOTE: If a cell is currently functioning properly, any change will most likely result in a problem with that functioning.) If the cell survives the interaction, it will most likely be the result of a mutation that makes it different than the surrounding cells and thus, in all likelihood, a competitor. If the new anomalous cell is better at surviving than the surrounding cells, it will crowd them out, becoming what is known as cancerous.

Meiosis

In meiotic chromosomal interactions, the danger is that negative interactions may be passed on to future generations. The mutations result in a variety of well-known conditions. The future may hold more or fewer, and it is anybody's guess.

A. Epistasis

When one gene masks the expression of another gene (e.g., if the first gene codes for the absence of a trait), as in no pigmentation, then the expression of the second gene, which may code for the kind of the pigmentation, has no effect.

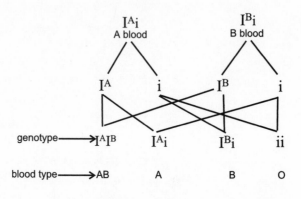

Hybrid A and Hybrid B Cross

B. Multiple alleles

Expression of a trait is controlled by the presence of more than two alleles, as in the case of blood types A, B, AB, and O. The alleles for blood types A and B (represented by I^A and I^B, respectively) are separately dominant. Together they are incompletely dominant as in AB (represented only by $I^A I^B$) blood type. Blood type O (represented by i) is recessive. If the two alleles are $I^A I^A$ or $I^A i$, then the dominate blood type will be A. If the two alleles are $I^B I^B$ or $I^B i$, then the dominate blood type will be B. Type O blood can only be ii. The interactions of these alleles, which produce an easily identifiable trait, is the first line of investigation in verifying the parents of a baby, if it actually comes into question, as has occurred in the past in hospital mix-ups.

C. Pleiotropy

Occurs when a gene has more than one phenotypic expression.

D. Polygenic inheritance

Rather than producing clearly defined phenotype, polygenic inheritance involves a continuum of genetic variation that is controlled by many genes, as the name suggests.

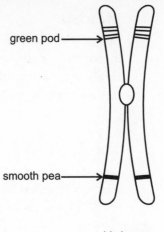

Linkage

E. Linked genes

If two genes are on the same chromosome, they linked. The mapping of (or the percent probability of spacing between and even the sequence of) these genes on the chromosome can be determined through the frequency with which crossing over occurs. One well-known instance of linkage is the universe of traits that is known as sex-linked. While these show up, as a rule, in a higher percentage of males, the sex-linked trait resides on the sex cell (gamete). In essence, the gene on the X chromosome has no counterpart on the male or Y chromosome, making the gene on the X chromosome dominant by default. Sex linkage is illustrated by, for example, as in the case of hemophilia, X^N where the N represents the allele for normal blood and X^n would represent an X chromosome with the n allele for hemophilia. Thus $X^N X^N$ and $X^N X^n$ would represent two females with normal blood, and the latter would be known as a carrier, able to pass on the gene but not expressing it as a phenotype. A female with the genotype $X^n X^n$ would be a hemophiliac. A male with the genotype $X^N Y$ would be expressing a normal blood phenotype, and a male inheriting $X^n Y$ would be hemophiliac, which is a rare condition. Several Punnett squares illustrating these interactions are noted below. Note that females are likely to be color blind 33 percent of the time while males are likely to be color blind 50 percent of the time in these interactions. This is also illustrated on the following page.

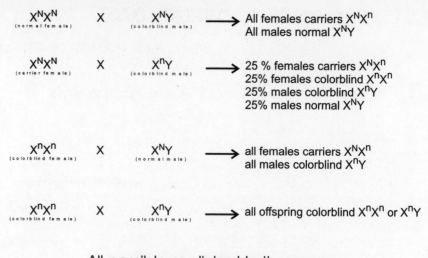

$X^N X^N$ (normal female) X $X^N Y$ (colorblind male) → All females carriers $X^N X^n$
All males normal $X^N Y$

$X^N X^n$ (carrier female) X $X^n Y$ (colorblind male) → 25 % females carriers $X^N X^n$
25% females colorblind $X^n X^n$
25% males colorblind $X^n Y$
25% males normal $X^N Y$

$X^n X^n$ (colorblind female) X $X^N Y$ (normal male) → all females carriers $X^N X^n$
all males colorblind $X^n Y$

$X^n X^n$ (colorblind female) X $X^n Y$ (colorblind male) → all offspring colorblind $X^n X^n$ or $X^n Y$

All possible sex-linked trait crosses
except color-blindness

F. Nondisjunction

Occurs when chromosomes do not properly separate during meiosis as they migrate to opposite poles. Down syndrome or Trisomy 21 is an example of this and is easily identified through amniocentesis, an examination of the baby's cells—the only ones present—in the amniotic fluid.

MULTIPLE CHOICE QUESTIONS

1. Which of the following is NOT indicated by Mendel's experiments?

   (A) incomplete dominance
   (B) segregation
   (C) recessive
   (D) dominant
   (E) independent assortment

2. A ratio of 3:1 in the phenotype of an organism is occurs when

   (A) the alleles are incompletely dominant
   (B) only recessive traits are involved
   (C) only dominant traits are involved
   (D) alleles segregate during meiosis
   (E) crossing over has occurred in Anaphase II

3. Color blindness in humans is

   (A) caused by a recessive allele
   (B) in equal proportion in both sexes
   (C) caused in females by a heterozygous genotype
   (D) inherited by males from their male parent
   (E) caused in males by a homozygous genotype

4. Trisomy 21 in humans is the result of

   (A) pleiotropy
   (B) polygenic inheritance
   (C) epistasis
   (D) x-inactivation
   (E) nondisjunction

5. Homologous chromosomes line up in pairs in

   (A) metaphase of mitosis
   (B) metaphase I
   (C) metaphase II
   (D) interphase
   (E) prophase of mitosis

6. In a heterozygous monohybrid cross, the dominant trait can be expressed in the phenotype of the F1 _____ of the time.

    (A) 0 percent

    (B) 25 percent

    (C) 33 percent

    (D) 75 percent

    (E) 100 percent

7. Which of the following would be the result of a true dihybrid cross?

    (A) AABB

    (B) AABb

    (C) AaBb

    (D) AaBB

    (E) aaBb

8. Which of the following would indicate a test cross, where **T** represents the dominant, Tall, trait and **t** represents the recessive, short, trait?

    (A) Tall x TT

    (B) Tall x Tt

    (C) Tall x tt

    (D) TT x Tt

    (E) short x tt

9. _____ refers to one gene affecting many traits.

    (A) Polygenesis

    (B) Pleiotropy

    (C) Linkage

    (D) Epistasis

    (E) Nondisjunction

10. Which of the following blood types are possible if the parents are A and O blood types?

(A) A and O

(B) B and O

(C) AB only

(D) O only

(E) A, B, and O

EXPLANATION OF ANSWERS FOR MULTIPLE CHOICE QUESTIONS

1. **The correct answer is (A).** The only choice that represents non-Mendelian genetics would be choice (A). The interaction of traits that blend would have been much harder to delineate without first knowing the laws that Mendel proposed. Being the mathematician that he was, Mendel reduced the number of variables to easily interpretable parameters. Choices (B) through (E), of course, reflect his laws: choice (B) represents the Law of Segregation, choices (C) and (D) represent the Law of Dominance, and choice (E) represents the Law of Independent Assortment.

2. **The correct answer is (D).** A phenotypic ratio of 3:1 is expressed only where there is the presence of a dominant allele in both individuals and a recessive allele in both individuals; they are hybrids. More important, though, is the fact that these alleles, in order to produce all possible combinations, must segregate in a process we now know as meiosis. This results in offspring with 1-TT genotype, 2-Tt genotypes, and 1-tt genotype. It produces a phenotype ratio of three dominant-expressing individuals and one recessive-expressing individual for every four offspring. A 3:1 ratio cannot occur in incomplete dominance. If only recessive or only dominant alleles are involved, the ratios would respectively be 100 percent and 100 percent. Crossing over cannot be detected using this ratio.

3. **The correct answer is (A).** Color blindness is caused by a recessive allele on the X chromosome. It is not found in equal proportion in both sexes: Females are 33 percent likely to be color blind and must inherit a color-blind–carrying X chromosome from both parents, but males are 50 percent likely to be color blind as they only need to inherit one color-blind–carrying X chromosome from their mother. A heterozygous female will be a carrier; she possesses one of the genes and has no color blindness, but in order to be color blind, she must be homozygous for the recessive trait. Since the male inherits only a Y chromosome from his father, he cannot inherit color blindness

from his father. Finally, males do not have homozygous sex chromosomes; therefore, color blindness is not a male-only trait. The homozygous genotype would only be in the female.

4. **The correct answer is (E).** Pleiotropy is one gene controlling many traits, and polygenic inheritance is the opposite of this. Epistasis is one gene affecting one other gene in its expression. X-inactivation occurs in mammals when one X chromosome stays coiled up and is known as a Barr body. The other one functions as the active chromosome. The coiled up autosome has become inactive. Trisomy 21 occurs when the twenty-first pair of chromosomes separate incorrectly during meiosis and is known as nondisjunction.

5. **The correct answer is (A).** In meiosis, the homologous pairs line up as individuals in meiosis I in preparation for final segregation in meiosis II. Meiosis simply needs to ensure the pairs end up in opposite daughter cells in meiosis I. It is in mitosis that the homologous pairs line up as pairs, and interphase, of course, is not a phase where the chromosomes even appear.

6. **The correct answer is (D).** In a heterozygous cross, height, for example, when **T** represents the dominant, tall, trait and **t** represents the recessive, short, trait, the following genotype would represent both parents: **Tt**. Following Mendelian procedures, and using a Punnett square to ensure accuracy, the following genotypes would result: **TT**, **Tt**, **Tt**, and **tt**. This would produce an occurrence of the dominant trait 75 percent of the time.

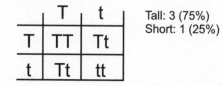

7. **The correct answer is (C).** The use of the word "true" dihybrid indicates that the test maker wants hybrid expression for both traits, which means the presence of all possible alleles in the genotype; this only appears in choice (C). Choices (B) and (D) are dihybrids, but not true dihybrids for both traits.

8. **The correct answer is (C).** A test cross occurs where a recessively expressed phenotype (which is therefore homozygous in genotype) is used to detect the likelihood of a recessive allele being masked by a dominant allele. If the recessive is crossed with a pure dominant, there is no chance of the recessive trait in the phenotype of the F1. If, on the other hand, there is a recessive allele being masked by a dominant allele, it has the

potential of showing up no less than 50 percent of the time in the phenotype of the F1.

9. **The correct answer is (B).** Our definition of one gene affecting many traits is, of course, called pleiotropy and occurs in instances such as several disease-causing alleles. Sickle-cell anemia is one such condition. The "sickling" or collapse of the RBCs results in effects in other parts of the body. Polygenesis is the opposite and would be found in such instances as height in humans, where a group of genes result in a continuum of heights. Epistasis is one gene affecting one other gene in its expression. Nondisjunction occurs when a pair of chromosomes separate incorrectly during meiosis.

10. **The correct answer is (E).** This is a case of multiple alleles controlling a trait (not to be confused with polygenesis). The alleles for blood types A and B (represented by I^A and I^B, respectively) are separately dominant, when one is present but the other is not. Taken together, they are incompletely dominant as in AB (represented only by $I^A I^B$) blood type. Blood type O (represented by i) is recessive. If the two alleles are $I^A I^A$ or $I^A i$, then the blood type will be A. If the two alleles are $I^B I^B$ or $I^B i$, then the blood type will be B. Type O blood can only be ii. In this case, the parents could be $I^A i$ and $I^B i$, which will produce all blood types.

VOCABULARY

alleles
autosomes
carrier
chromatin
codominance
color blindness
crossing over
dihybrid cross
diploid cell
dominant
F2 generation
filial (F) generation
gametes
gametogenesis
genes
genotype
germ cells
gonads
Gregor Mendel
haploid cell
hemophilia
heterozygous
homologous chromosomes
homozygous
incomplete dominance

law of dominance
law of independent assortment
law of segregation
meiosis
monohybrid cross
oogenesis
ovaries
parent (P1) generation
phenotype
polygenic inheritance
Punnett square
recessive
sex cells
sex chromosomes
sex-linked traits
sister chromatids
spermatogenesis
spindle fibers
synapsis
test cross
testes
tetrad
trait

Chapter 5

MOLECULAR GENETICS: DNA AND EVOLUTION

OVERVIEW

DNA controls the life of the cell by controlling the manufacture of proteins and is the material of the genes that are integral parts of the chromosomes. These proteins are used to either make other molecules or form part of the structural network of the cell. DNA even makes the proteins that control the functions of DNA itself as well as controlling heredity. Watson and Crick are credited with first expressing the structure of DNA. The student will find questions on both the structure and function of DNA as well as RNA and the molecules involved in protein synthesis. DNA is now known to be a double helix in its non-replicating state.

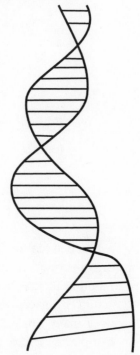

DNA double helix

The code for life resides in the sequence of nucleotides along the length of the molecule. The foundation of evolution is change that, of course, comes about through an organism's unique DNA of the cell as it impacts the organism's interactions with its surroundings. For this reason, we include material on evolution coupled with the material on DNA. Since the function of DNA is to pass on traits from generation to generation, and we have explored heredity, we now explore the *details* of how this information is passed on. As we have stated before, the student should be familiar with the material outlined in this book and related material as well. We begin with the structure of DNA.

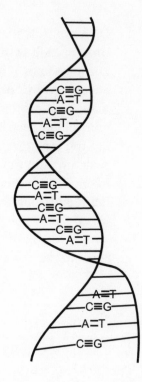

DNA: sample base-pairing

DNA

A biochemical molecule responsible for passing on the traits of an organism, which include controlling the biochemical nature and function of that organism. DNA is a double helix molecule that achieves a great deal of compacting in the nucleus by coiling in on itself many times. DNA makes RNA, which carries the code to the cytoplasm. There, ribosomes are utilized in making polypeptide chains that are proteins. The lack of an oxygen on the ribose sugar—thus the name deoxyribose nucleic acid—and the presence of the base thymine both serve to prevent DNA from leaving the nucleus.

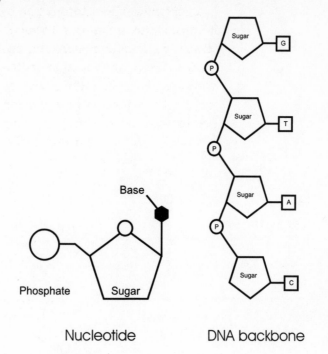

Nucleotide   DNA backbone

NUCLEOTIDES

Are the building blocks of DNA, they are used to make up the repeating units in the strands of DNA that represent the genetic code. A nucleotide consists of a sugar, a phosphate, and a nitrogen base. Since there are only four bases, researchers postulated that the enormous amount of genetic variation on the planet had to be in the sequence of the nucleotides within the DNA molecule. This sequence then controls the synthesis of precise proteins in the sequence of amino acids. On one end of the DNA molecule, the 5-carbon sugar has a phosphate attached and is known as the 5′ end. On the other end of the sugar is an OH that is identified as the 3′ end of the DNA molecule.

Bases

Four nitrogenous bases make up the bonding sites along the center of the DNA molecule and are bonded to a carbon on the sugar molecule. There are two purines and two pyrimidines. The purine names end in *-nine*.

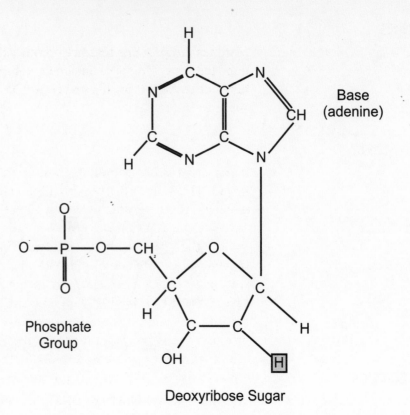

Deoxyribose Sugar

DNA Nucleotide

A. Adenine

A purine, double-ring base with two bonding sites.

B. Thymine

A pyrimidine, single-ring base with two bonding sites.

C. Cytosine

A. pyrimidine, single-ring base with three bonding sites.

D. Guanine

A purine, double-ring base with three bonding sites.

Ribose

A 5-carbon sugar—signified by the *-ose* ending—it is considered the central part of the nucleotide, as the bases and the phosphate bond to it.

Phosphate

Bonded to another carbon on the ribose sugar, the last phosphate on the molecule is the 5′ end.

HISTONES

Histones are proteins that help protect the DNA molecule in the uncoiled state. The DNA molecule is surrounded by eight or nine histones to help it form a protective DNA-histone complex in a tight space in the nucleus.

BASE-PAIRING

Base-pairing is the pairing of complimentary bases along the DNA strand. The sum of the bonds and the coiling of DNA makes the molecule securely attached along its entire length. Analysis of the assays of the DNA from a variety of organisms caused Erwin Chargaff to note that the percentage of adenine was almost identical to the percentage of thymine in the DNA of a cell. A similar relationship was discovered for cytosine and guanine. This was later referred to as Chargaff's rule and led Watson and Crick to the ultimate conclusion that the molecule was directed inward—the result of which was their proposal that DNA was a double helix. Linus Pauling, who did so much work with proteins, wrestled with the molecule being directed outward as did others, many on the strength of Dr. Pauling's reputation.

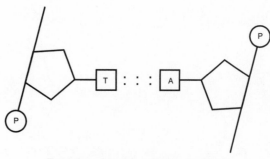

Thymine-Adenine base-pairing

Adenine-Thymine

This purine-pyrimidine bonding is result of each base having two hydrogen bonding sites. Adenine and thymine can bond only with each other.

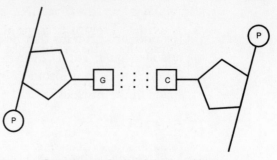

Guanine-cytosine base-pairing

Cytosine-Guanine

This purine-pyrimidine bonding in the DNA molecule is the result of these bases having three sites for hydrogen bonds. Cytosine and Guanine can bond with each other only.

H+ bonds

The bonds between these two classes of chemicals that form the base sequences in DNA, in addition to other places where we find hydrogen bonds.

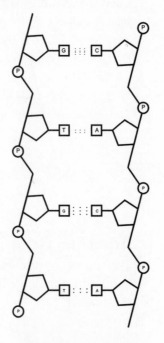

DNA base-paring

COMPLIMENTARY STRANDS

Since adenine always bonds with thymine and guanine bonds with cytosine, if one strand of the DNA molecule is known, then the other, complimentary strand can be known. This means that if you have one half of the molecule, you can construct the other half,

Peterson's AP Success: Biology

which is exactly what DNA does during replication. The main strand is used as a template to produce its complement. The nature of the bonds along the strand make the adenines on one strand line up in the opposite direction than the adenines do on the other strand. As a result, the molecule is said to be anti-parallel.

5′ end

"Hanging out" on one end of the base pairs is the phosphate group, which is the end that starts the "reading" of the molecule when it is being replicated. The direction of replication along the master strand then is from 5′ end to 3′ end.

3′ end

opposite the base on the master strand with its 5′ end is the complimentary strand with its 3′ end oriented outward at the "beginning" of this strand, in an anti-parallel way. (See the following diagram)

REPLICATION

A combination of the words "*rep*roduce" and "dup*licate*," this refers to the act of DNA making a copy of itself. This precedes mitosis or meiosis. Mitigated by enzymes, it proceeds as two concurrent events, one from the 5′ end of what is called the leading strand and the other from the 3′ end of the lagging strand. The result of replication is said to be semi-conservative, since we end up with half the original DNA in each of the resulting new strands.

DNA helicase

DNA helicase is an enzyme that begins the unraveling of the DNA molecule at the sites of the hydrogen bonds.

DNA polymerase

This enzyme arranges the new nucleotides next to their complimentary base to make the new strand of DNA. As the name suggests, it makes a polymer out of individual nucleotides.

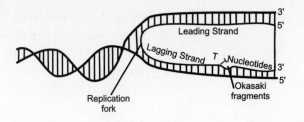

DNA replication

Leading strand

A strand of DNA that starts at the 5' end and is made continuously. Not named for the fact that it starts first, but for the fact that, since it is made continuously, the construction of it proceeds faster.

Lagging strand

A strand of DNA that starts at the 3' end. Its production proceeds slower than the leading strand because it is made in pieces that are then bonded to the template to which it will be complimentary. These pieces, known as Okasaki fragments, are bonded into place by DNA ligases.

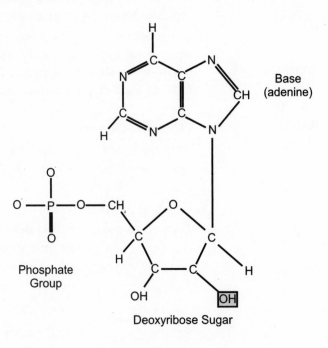

DNA nucleotide

RNA-RNA

RNA-RNA is a 5-carbon sugar that possesses an extra oxygen atom [compared with DNA] and replaces thymine with the base uracil, thus enabling it to pass through the nuclear envelope and take the code of DNA to the cytoplasm. Three types of RNA are made in the nucleus and reunite in the cytoplasm in the process known as protein synthesis.

TYPES

RNA, the carrier of the DNA code from the nucleus to the RER in the cytoplasm, has three forms, each playing a key role in the synthesis of proteins. These molecules, acting in concert, ultimately produce the proteins that control the life of the cell, even the production of RNA.

rRNA

rRNA stored in the nucleolus, it helps make up ribosomes which reside on the RER. Ribosomes are composed of rRNA and proteins. The mRNA attaches to the ribosomes and thus begins the making of a protein.

mRNA

mRNA is the lengthy form of RNA that is made of DNA, the one that codes for the protein to be made in the RER. This sequence will be "read" in the ribosome and serves as the blueprint for the precise sequence of amino acids that will make up the protein coded for in the mRNA. The codon is three consecutive nucleotides on the mRNA that code for a particular amino acid carried by the tRNA.

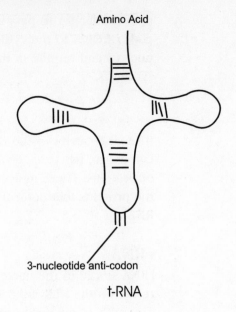

t-RNA

tRNA

tRNA contains the anticodon to the mRNA's codon, just three nucleotides long; the nature of this molecule was hypothesized before the structure was actually known. The tRNA carries an amino acid to the ribosome, where it bonds to the codon on the mRNA. There are 20 amino acids that make up proteins. The sequence of these amino acids, like the sequence of nucleotide bases in the DNA, is critical. If there was one base pair for every one amino acid, this would result in only four amino acids ever being utilized to make proteins. Two base pairs for every amino acid will code for a maximum of 16 amino acids, four short of the needed 20 to transport all of the amino acids needed for life. It is now known that the tRNA molecules are, in fact, three bases long, providing more than enough variations to code for 20 amino acids as well as stop, start, and some duplication.

First letter	Second Letter				Third Letter
	U	C	A	G	
U	phenylatanine phenylatanine leuche leuche	serine serine serine serine	tyroshe tyroshe STOP STOP	cysteine cysteine STOP tryptophan	U C A G
C	leuche leuche leuche leuche	proline proline proline proline	histidine histidine glutamine glutamine	arginine arginine arginine arginine	U C A G
A	isoleuche isoleuche isoleuche methionine & START	theonine theonine theonine theonine	asparagine asparagine lysine lysine	serine serine arginine arginine	U C A G
G	valine valine valine valine	alanine alanine alanine alanine	aspartate aspartate glutamate glutamate	glycine glycine glycine glycine	U C A G

mRMA codons and accompanying amino acids

TRANSCRIPTION

Transcription takes place in the nucleus of eukaryotic cells. It is the first step in protein synthesis, following replication, wherein DNA makes RNA. Three RNA molecules, rRNA, mRNA, and tRNA are made during this phase, each from the separate complementary strands of original DNA. They are then transported to the cytoplasm where the next step is performed: translation. During translation, the sequence of codons on mRNA orders the sequence of amino acids in the protein. Transcription can be likened to what occurs in, say, a court setting, where the court reporter transcribes the spoken word into the written word: same language, different form.

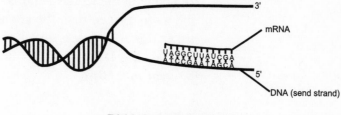

mRNA transcription

In the nucleus, a similar nucleic acid making a nucleic acid event occurs. This time it is when DNA makes RNA: same language—nucleotides—different form—DNA and RNA.

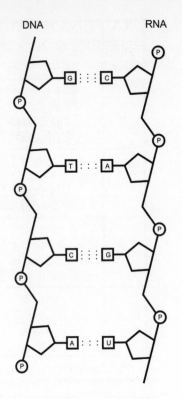

DNA-RNA base-pairing

Bases

In RNA, uracil replaces the thymine found in DNA. The uracil, like the thymine it replaces, still bonds opposite adenine. Students should be careful not to associate uracil with thymine in a way that the uracil *replaces* the adenine. It doesn't; like thymine, it will bond *opposite* adenine as thymine does in DNA. In RNA, we will get A-U pairing, whereas in DNA, we get A-T pairing.

Sugar

The 5-carbon sugar that, along with the phosphate, is part of the backbone of nucleotides and is different in RNA. The RNA sugar has one more oxygen than the amount found in the DNA sugar. This makes this site possess an OH radical with no electrical activity. Alternatively, DNA possesses an H+ at that site, making its activity different than RNA.

Promoters

The site on DNA where transcription of RNA is begun, using only one of the DNA strands, the sense strand. The other strand—called the anti-sense strand—is not used during this process.

RNA Polymerase

As the name of this enzyme suggests, its role is to bring RNA nucleotides into proper position on the sense strand ending in an mRNA strand.

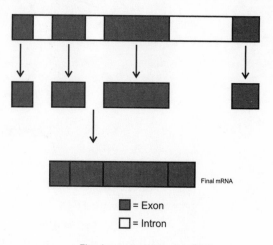

■ = Exon

□ = Intron

Final processing of mRNA

RNA modification

The mRNA molecule is not ready to be transported to the cytoplasm yet. It undergoes a final processing stage where the nonsense sequences, called introns, are excised, leaving the meaningful sequences known as exons to make up the final mRNA. A tail, called the poly(A) tail, is added to the 3′ end, a cap—5′ cap—is added to the 5′ end, and the molecule is ready to take part in the making of polypeptides.

TRANSLATION

Continuing the court reporter analogy, this stage of protein synthesis is like translating words in one language into another language. In the case of the cell, we will be translating from the language of the nucleus into the language of the cytoplasm or cell on a larger scale. DNA has been replicated, it has transcribed its message into RNA, and a nucleic acid that can safely leave the nucleus without having the original code is being compromised if anything should happen to this RNA. Finally, we translate the nucleotide sequences into polypeptide sequences.

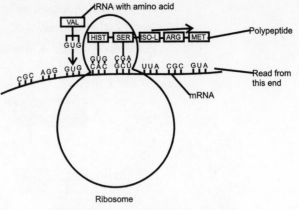

Translation of RNA into proteins

mRNA

This is known as the codon and reflects the statement of the original code in the DNA. This will be placed along the ribosomes for "reading," the sequential passing of the base sequence over ribosome and subsequent bonding of tRNAs in proper places on the mRNA.

tRNA–tRNA molecules

tRNA–tRNA molecules transfer amino acids to the ribosomes. There is at least one tRNA for each of the 20 amino acids. Amino acids bond to one end of the tRNA. The other end contains a three-base antic-odon sequence form of RNA, which codes for specific amino acids. While their nucleotide end anticodon bonds with the next sequence on the mRNA, their amino acid form lines up with the next amino acid.

A. Start

The mRNA sequence that initiates the construction of all polypeptides is A-U-G, which is bonded with the tRNA sequence of U-A-C that carries the amino acid methionine.

B. Stop

There are several stop codons on the mRNA, namely U-A-A, U-A-G, and U-G-A, which bond opposite the tRNA anticodons of A-U-U, A-U-C, and A-C-U, respectively.

Polypeptides

The chains of amino acids that make up the polypeptides are, in sum, formed in the following way. Once the mRNA is attached to the rRNA in the ribosomes, the tRNA anticodon, with its accompanying amino acids, moves into position on the codon of mRNA. The mRNA is then shifted by three bases along the rRNA and makes room for another tRNA-amino acid complex to move into position. Once this occurs, the two adjacent amino acids are bonded to each other in a peptide bond at the amine and the carboxyl groups. Once again, mRNA is shifted, and another tRNA-amino acid moves into place. Another peptide bond is formed between this third amino acid in the sequence and the second amino acid. As the process continues, the tRNAs are immediately rejected from the mRNA—the tRNA-amino acid complex is the substance that was moved into position on the mRNA, NOT the tRNA by itself. It is now available to leave the area and pick up another amino acid specific to itself. In addition, the molecule is continually moved along the rRNA sites, and, thus the polypeptide elongates. When codons U-A-A, U-A-G, or U-G-A are encountered, the process stops and a polypeptide is formed. The formation of a protein may not be complete and depends on the interaction among the amino acids as to what form this protein will take. Some, like insulin, are actually shortened in order to become active. Not all proteins will become enzymes, some will be structural as in the proteins in the plasma membrane or those that make up part of other biological molecules. Originally, the thinking was characterized as "one gene—one enzyme," but is now thought to be more accurate with the phrase "one gene—one polypeptide chain."

MUTATIONS

Characterized as "errors" in the genetic code, these cause a change in the functioning of the cell, either as a structural anomaly—as in the case of sickle cell anemia—or as a functioning anomaly, such as a defective enzyme. There are so-called micro-mutations and macro-mutations. Both form the basis for much of our evidence for evolution. The causes of mutations can be almost anything from radiation, to physical disruption of the cell—a rare one—to environmental factors to chemicals. Pollution is feared to be the potential cause of many mutations.

POINT

Point mutations are those that occur at a specific site along the DNA molecule, specifically a change in a single nucleotide.

Base substitution
One base along the original DNA sequence is replaced by another base.

Insertion
A base is added into the original DNA sequence.

Deletion
A base is deleted from the original DNA sequence.

Frame shift
The entire sequence is shifted one or more frames along the DNA sequence.

MACRO-MUTATIONS

These mutations tend to affect larger numbers of DNA sequences, even up to whole chromosomes, as in the case of the first type discussed below.

Non-disjunction
Chromosomes fail to separate properly during the process of meiosis as in the cases of Down's, Turner's or Klinefelter's syndromes.

Translocation
A segment breaks off of one chromosome and moves to another chromosome.

Inversion
A segment of a chromosome breaks off and is inserted in reverse order.

Deletion
A segment of a chromosome breaks off and is lost.

Duplication
An extra copy of a segment of a chromosome is produced along the chromosome.

RECOMBINANT DNA

The transferring of DNA segments from one entity to another, whether it is DNA molecules or chromosomes. Recent technology uses restriction enzymes from bacteria to slice DNA in very specific places for the purposes of recombining sequences or assaying the nature of a sequence.

BACTERIAL GENETICS

Variation is introduced into bacteria cells in several ways. One is by conjugation, the exchange of DNA between bacteria. Another, transduction, occurs when DNA is introduced into the bacteria by a virus, and transformation occurs when bacteria absorb free pieces of DNA from their environment.

VIRAL GENETICS

Non-living substances, viruses can nonetheless take over a living cell in two stages. The first, called the lytic stage, is when the virus penetrates the cell, uses cellular enzymes to duplicate viral particles, transcribes the DNA into RNA, and then uses the RNA to make proteins. The second, called the lysogenic stage, occurs when an infected bacterium does not immediately duplicate viral particles. The viral DNA is temporarily incorporated into the cellular DNA.

EVOLUTION

Changes in population over a period of time, usually millions of years; 8 percent of the AP exam has questions on this subject. Much of what involves evolution also involves knowledge of several fields of geology, as will be seen. We will provide the highlights of this concept and make reference to areas that the student will want to be conversant with outside the specific area of biology, such as paleobiology, biogeography, etc.

JEAN-BAPTISTE DE LAMARCK

Lamarck popularized the notion that acquired traits were passed on to future generations. Changes in the body cells are not capable of being passed on. Only sex cell changes achieve this. Lamarck's theory became known as the use-disuse theory and was soon seen by many as fruitless. If one loses a finger prior to having children, those children will not be missing a finger as a result of the parent losing a finger. Especially with our knowledge of the nature of mutations, we can reject this idea outright. Darwin did also.

CHARLES DARWIN

The 19th-century naturalist who advanced the notion of Natural Selection, which became the foundation of his theory that led him to conclude how species came to be so varied on the planet. Basically, he stated that nature selects those organisms that will survive and passes on their survivability traits.

Natural Selection

Darwin's idea that species could not simply acquire the trait that allowed them to survive. Like the culling of a herd that predators achieve, the ability to survive better than other members of the species led him to conclude that nature was bringing about the changes—we now know the level on which this proceeds. Changes in DNA, and/or DNA combinations provide the variation; nature selects the fittest. Darwin said it as follows.

A. Number of offspring

Species produce more offspring than the environment can support.

B. Competition

As a result of overproduction, these offspring are led to compete with each other for survival: food, shelter, water, etc.

C. Variation

Within every population, there is the potential of having variation among the members. Students will note that this is the place that modern molecular genetics has provided evidence for this to happen.

D. Survival of the fittest

The more fit in the population will survive to mate.

E. Evidence

Evidence comes from a variety of scientific areas of study, as was mentioned before. They are, but are not limited to: biogeography, comparative anatomy, embryology, molecular biology (which is the main subject of this chapter) and paleontology.

F. Types

There are three ways in which nature "selects" the fittest to survive, and they have significance in terms of how evolution will proceed.

1. *Directional*—phenotypes that are at the extremes of a range of traits are selected against those at the opposite end of the range.

2. *Stabilizing*—selects out individuals with extreme traits, favoring those with more common traits. Those with the extreme traits have fewer means to survive.

3. *Disruptive*—favors the extremes mainly when the environmental changes favor these extreme traits.

4. *Sexual*—differential mating of a sex in a species that ensures their survivability. For example, if a female mates with a superior male and produces superior offspring, she will more likely be able to survive now through that male's protective attachment.

5. *Artificial*—directional selection mediated by humans.

Speciation

Speciation is the process by which a new species originates. The emergence of a new species as a result of the interactions within molecular and environmental parameters producing what is known as divergent evolution: two closely related species emerge from a common predecessor. Members of the same species look alike, behave alike, and can mate and produce offspring that can mate.

A. Types

There are two known types of speciation that center around the variation that can occur when a barrier arises between members of a species or when no barrier arises, species simply diverge on their own—caused by the "usual" factors, of course.

1. *Allopatric*—a geographical barrier separates parts of a population and, over time, the two diverge to become two distinctly different species.

2. *Sympatric*—a geographical barrier does NOT separate members of a population yet, over time, the two diverge to become two distinctly different species. This occurs often in the plant kingdom.

B. Sources

Speciation of variation within a population can be caused by a variety of factors acting alone or in concert.

1. *Mutations*—many mutations or mistakes in the DNA makeup are lethal and do not get passed on. Alleles never before in the population can arise, however, by mutations to the gene pool.

2. *Diploidy*—the production of two copies of each chromosome in a cell.

3. *Sexual reproduction*—during meiosis, events can emerge, such as independent assortment, random gamete selection, and crossing over.

4. *Outbreeding*—mating that occurs between unrelated partners.

5. *Balanced polymorphism*—the maintenance of a variety of phenotypes in a population, the most advantageous of which possess survival value. The non-advantageous alleles decrease in frequency.

C. Allele Frequency

Darwin proposed natural selection as the means by which evolution proceeded. With the advent of modern genetics, we have discovered other factors that obviously impact the nature and direction of evolution. Basically, these factors change the frequency of allele availability and thus have the effect of advancing evolution. Besides mutations and natural selection itself, two factors already discussed, are briefly outlined below.

1. *Gene flow*—changes that occur in the population when individuals enter or leave the population.

2. *Genetic drift*—random increase or decrease in alleles.

3. *Nonrandom mating*—when individuals choose with whom to mate, including inbreeding—the mating with relatives—and sexual selection—the selection by females of a particular male.

D. Genetic equilibrium

Genetic equilibrium, also called the Hardy-Weinberg equilibrium, this is a favorite subject of AP Biology test-makers. Genetic equilibrium occurs when allele frequency remains unchanged—meaning, there is no evolution: factors—mentioned above as affecting allele frequency—are held constant. The frequency for each allele is given by p and q, the frequency of homozygous individuals is given by p^2 and q^2, and the frequency of heterozygous individuals is given by $pq + qp = 2pq$. The following equations hold: (1) $p + q = 1$ (the sum of the alleles is 100 percent) and (2) $p^2 + 2pq + q^2 = 1$ (the sum of all the individuals is 100 percent). This equation holds true if

1. no natural selection occurs

2. no mutations occur

3. the population is isolated from other populations

4. the population is large

5. mating is random

Patterns of evolution

Four patterns emerge in evolutionary studies.

A. Convergent evolution

When two unrelated species share similar traits through adaptation to similar environmental conditions.

B. Divergent evolution

Allopatric or sympatric speciation lead to the emergence of two species from a common ancestor.

Peterson's AP Success: Biology

C. Parallel evolution

After diverging from a common ancestor, two unrelated species make similar evolutionary changes.

D. Co-evolution

An almost leap-frogging of traits where one species adapts to a new set of conditions and another species adapts to new conditions brought about by the new species.

MULTIPLE CHOICE QUESTIONS

1. Which of the bonding examples below is NOT possible?

   (A) a DNA adenine to a DNA thymine

   (B) a DNA adenine to an RNA thymine

   (C) a DNA guanine to an RNA cytosine

   (D) a DNA adenine to an RNA uracil

   (E) a DNA guanine to a DNA cytosine

2. Which of the following replaces guanine in RNA?

   (A) adenine

   (B) thymine

   (C) cytosine

   (D) uracil

   (E) none of the above

3. Which of the following statements is true about mutations?

   (A) Rates tend to be very high in most populations.

   (B) generally lethal

   (C) irreversible

   (D) Only certain gene locations are affected.

   (E) source of genetic variation

4. For the DNA strand 5′-A-C-C-G-T-G-A-C-A-T-T-G-3′, the correct compliment DNA would be:

   (A) 3′-T-G-G-C-A-C-T-G-T-A-A-C-5′

   (B) 5′-T-G-G-C-A-C-T-G-T-A-A-C-3′

   (C) 3′-U-G-G-C-A-G-U-G-U-A-A-C-5′

   (D) 5′-A-C-C-G-U-G-A-C-A-U-U-G-3′

   (E) 3′-T-C-C-G-A-G-T-G-T-A-A-C-5′

5. The portion of the DNA molecule that can vary is its

(A) sugar

(B) base

(C) deoxyribose

(D) ribose

(E) phosphate

6. An mRNA is 429 nucleotides long. The number of amino acids in the polypeptide chain formed from this mRNA is

(A) 143

(B) 142

(C) 141

(D) 429

(E) 428

7. Which of the following is NOT consistent with Griffith's experiments?

(A) injected mixture of R-strain and live S-strain: mouse dies

(B) injected mixture of heat-killed S-strain and live R-strain: mouse lives

(C) injected heat-killed S-strain: mouse lives

(D) injected R-strain: mouse lives

(E) injected S-strain: mouse dies

8. Which of the following is a semi-conservative process?

(A) DNA replication

(B) non-disjunction

(C) translation

(D) transcription

(E) crossing over

9. In the illustration below, the strand with an (*) is the

(A) promoter

(B) Okasaki fragment

(C) template

(D) lagging strand

(E) leading strand

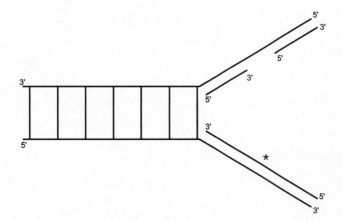

10. Darwin's Natural selection includes all of the following except

(A) Variation can be due to mutations.

(B) The difference in survivability between organisms may be due to variation.

(C) Not all organisms may survive, due to competition.

(D) Organisms produce more offspring than can survive.

(E) Some organisms will be more fit to survive than others.

EXPLANATION OF MULTIPLE CHOICE ANSWERS

1. **The correct answer is (B).** In RNA, thymine is replaced by uracil, so the correct choice is (B), where DNA adenine is said to bond to an RNA thymine. All the other choices are incorrect, including the DNA guanine to an RNA cytosine. The testmaker may depend on the student vaguely remembering that a certain base-pairing between DNA and RNA is not possible and may choose choice (C) as being correct. This is not the case.

2. **The correct answer is (E).** Guanine is not replaced in RNA. It is found in both DNA and RNA. The trick here, once again, is the vague recollection of something being replaced. Of course, thymine in DNA is replaced by uracil in RNA, and uracil is on the list. However, the question here asks about the replacement of guanine, which does not occur. The bases listed in choices (A) and (C) also are found in RNA.

3. **The correct answer is (C).** These recent conclusions about mutations—recall that Darwin did not know of mutations—are all the reverse of those listed in the choices, with the exception of choice (C), the correct answer. Rates, in fact, tend to be below in populations, mutations are generally not lethal, any gene location can be affected, and they are felt to be the source of genetic variation. Darwin felt over-production of offspring was the source of potential variation.

4. **The correct answer is (A).** The complimentary strand for the indicated DNA strand would begin at the 3' end, not at the 5' end. In addition, it would also have thymine (T). Finally, opposite (A) would be (T), and vice-versa, and opposite (G) would be (C), also vice-versa.

5. **The correct answer is (B).** Choices (A), (C), and (E) are all places on the DNA that are constant and form the sides of the DNA "ladder." The place at which the molecule differs is in the bases that form the rungs of this ladder, providing DNA with a virtually infinite number of variations. Ribose, of course, refers to RNA and should not be on the list when discussing DNA.

6. **The correct answer is (B).** tRNA, which contains 3 anti-codons bases for every amino acid it carries, matches up opposite the mRNA codons that are also three bases long. This means that the mRNA in question potentially codes for 429/3 amino acids. This gives us a potential of 143 amino acids or choice (A). However, the stop sequence does not code for an amino acid, so this gives us 142 amino acids coded for. Care should be exercised when considering the start sequence. While not a part of the resulting protein, that site does code for the amino acid methionine in the translation process.

7. **The correct answer is (B).** The foundation of Griffith's work was that the mice that received the killed Smooth-coat strain/ Rough-coat strain mix, in fact, died. This suggested that the inheritable trait of producing a lethal substance to mice could be inherited by the R-strain bacteria. All the other choices are true and led to the experiment outline briefly in choice (B), but the conclusion of that experiment, once again, was that all the mice died.

8. **The correct answer is (A).** In DNA replication, each new strand receives half of the original DNA, a process said to be semi-conservative—it conserves only half of the original DNA for each new strand produced. Crossing over and non-disjunction are macro-genetic events that occur during meiosis and pass on all of the genetic information for the event. Crossing over exchanges whole parts—however long—of one chromosome with whole

parts of a homologous chromosome. Non-disjunction causes an entire chromosome to be incorrectly separated. Translation and transcription pass along the entire code for a protein from RNA to polypeptide and DNA to RNA, respectively.

9. **The correct answer is (D).** The strand with a (*) is being replicated from the 5' end onward and will proceed without delay. The strand with the 3' end as a start—opposite the 5' end on its compliment, the sense strand—makes this strand the anti-sense strand and is fashioned piece by piece from Okasaki fragments. These fragments are bonded onto the anti-sense strand with DNA ligases. The construction of the 3' strand then proceeds slower than the construction of the 5' end and is said to lag behind. Since the strand with a (*) is the 5' end, it is not an Okasaki fragment, nor is it a template; the sense strand is the template and a promoter is a molecule that starts the process: it is not a DNA molecule, but a protein.

10. **The correct answer is (A).** Darwin did not know of mutations, so choice (A) is the correct answer. Fundamental to his idea of Natural Selection was that, in order, (1) organisms produce more offspring than can survive and, (2) due to competition, not all organisms will survive which means, (3) some organisms will be more fit to survive than others and (4) variation may be the reason for the difference in survivability.

VOCABULARY

adenine

anticodon

base pairing

base substitutions

codons

complementary

cytosine

deletions

deoxyribonucleic acid (DNA)

deoxyribose

DNA polymerase

DNA replication

double helix

Down's syndrome

duplication

exons

guanine

hydrogen bonds

insertions

introns

inversion

mRNA

mutation

nitrogenous base

nondisjunction

nucelotides

peptide bond

point mutation

promoters

protein synthesis

proteins

ribonucleic acid (RNA)

ribose

RNA polymerase

RNA primer

rRNA

template

thymine

transcription

translation

translocation

tRNA

uracil

Unit III— Organizations and Populations

Chapter 6

ORGANISMS AND POPULATIONS

OVERVIEW

This material constitutes about 50 percent of the AP Biology exam. This part of the study of biology, then, contains the bulk of the material on the scale of entire organisms. Previous chapters have dealt basically with how the organisms function mostly at the molecular level. We now explore organism-level structure—and, therefore, function—and on a grander scale, the biology of the organism and how organisms as populations interact with one another and with other organisms as well as their environment. Carolus Linnaeus is credited with organizing living things into understandable groups based on their structures, giving them a two-part name. Today, we use many characteristics, including behavior and, more recently, DNA analysis. Since earliest times, humans have attempted to understand their world, and one of the ways to start the process is to organize it; the way humans have done this has changed over time. This chapter is about the grouping of organisms. Some of the vocabulary at the end of the chapter will not be detailed here, and we suggest that the student know all of those terms; this holds true for all the chapters.

CLASSIFICATION

Taxonomy, or the science of classifying organisms, proceeds from the grouping of organisms into categories, based on various characteristics.

Schema

Just as we find in grocery stores, the organisms are placed in specific categories. Imagine trying to find an item in the store if all the delivery vehicles dumped their goods at the front door. In a like manner, grouping organisms has many benefits. Imagine, again, proposing that a predator of a parasite be introduced into the local environment in an attempt to control the parasite population, but instead, the organism you believe to be the predator is actually a slight variation of that predator and has no interest whatsoever in that parasite. You may, now, have two problems to deal with, and at the very least, you haven't done anything to help control the parasite.

Aristotle

Aristotle's classification system proposed that if something moves, it is an animal, and if it doesn't, it is a plant. Of course, sponges were mistakenly taken for plants, and when they threatened the shellfish industry centuries ago, were "killed" by being cut up and tossed back into the sea. To the surprise of the shellfish harvesters, the next year, the number of sponges had increased; the shellfishers were actually helping the sponges reproduce—asexually—by their actions.

Aristotle Taxonomy
Motile ... Animal
Non-Motile ... Plant

Linnaeus

Proposed naming organisms by a "two-name" system that we call binomial nomenclature. These were very specific names based on the organism's characteristics and are the genus and species of today. Note that the Genus is always capitalized and the species is not, as in Terrestris americanus, and is underlined.

Modern

The modern system of classification now contains five major groups called kingdoms. Life on the planet would be analogous to the entire grocery store—or any other store, for that matter. The major consumer item areas, such as produce, dairy, canned goods, etc., would be analogous to the kingdoms of living things. As with the grocery store, the sub-categories get more and more specific until it is possible to name an item exclusive of all other items in the store. The AP exam tests significantly in this area. The student should know the categories, starting with the largest, the kingdom, and continuing to narrower and narrower groups in the sequence: Kingdom, Phylum, Class, Order, Family, Genus, and Species. In plants, the word "phylum" is replaced with the word "division." Modern day study has become so detailed that we will now find such categories as Sub-genus, Super-species, etc. The student is responsible only for the above schema. Most beginning biology students become familiar with the mnemonic—memory device—in a variety of expressions, one being *King Phillip Come Out For Goodness Sakes*, the first letter of which, and in the order given, is the first letter of each of the major categories in taxonomy.

<u>Mondern Taxonomy</u>

Kingdom
Phylum
Class
Order
Family
Genus
Species

EXAMPLES OF GENUS AND SPECIES

Canis familiarus
domesticated dog

Felis feline
domesticated cats

Homo sapiens
modern humans

MODERN TAXONOMY

Today, scientists generally agree on five major categories of living things, or kingdoms. Students may want to become familiar with how these categories came to be agreed upon. Students should recall that some of these organisms were mentioned in other chapters and at other levels. The good news is that the ETS does not test for information as detailed as how the phylogenic categories arose. A brief word about viruses seems appropriate at the beginning of a discussion on the taxonomy of living things. Viruses, while not considered a living thing, can, nonetheless, take over the cell chemistry of a living thing to reproduce. We will now outline the schema of living things.

PROKARYOTE (MONERAN)

Single-celled, microscopic prokaryotic—see chapter 2—cells with no distinct nucleus or other membrane-enclosed organelles.

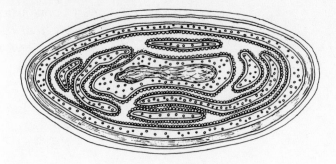

Typical Prokaryotic cell

Bacteria

Bacteria have cell walls composed of peptidoglycan, an amino acid-sugar complex, and circular DNA. Composition of the cell wall provides us with the Gram staining means of identifying certain types of bacteria. Reproduction of bacteria was outlined in an earlier chapter. Some bacteria possess flagella. While many bacteria are decomposers, some fix nitrogen and other elements in a form usable by organisms, and some are pathogenic.

A. Shapes

Bacteria can be found in three shapes: coccus (round-shaped), bacillus (rod-shaped)—the one the AP Biology exam commonly refers to when asking a question about bacteria—and spirillus (spiral-shaped).

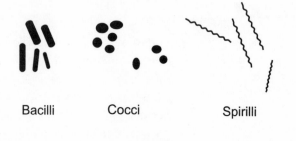

| Bacilli | Cocci | Spirilli |

Bacteria Types

B. Types

Aerobic (oxygen-needing) bacteria are the largest group of bacteria. Anaerobic (not needing oxygen) bacteria are found in two groups: those who need an oxygen-free environment (obligate) and those who do not need a lack of oxygen, but a small amount of oxygen will not kill them (facultative).

C. Significance
Bacteria are the most populous organism on earth—they can be useful in making pickles from cucumbers, or they can be lethal, as in the case of pneumococcus and several STDs.

Cyanobacteria
Producer prokaryotes, their numbers can "explode" under the right conditions and are the cause of water pollution associated with algal blooms.

PROTISTA

Protista (also called protoctist) single-celled eukaryotes, this kingdom is a composite for organisms that do not fit in other kingdoms.

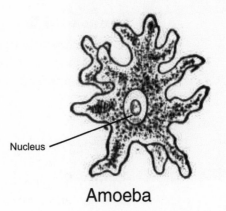

Nucleus

Amoeba

Protozoan
Protozoan are classified by their mode of locomotion. Amoeba and paramecium are two of the Protists about which test questions are asked. The amoeba is a heterotrophic, shapeless organism that moves by pseudopods and engulfs its food by the endocytotic method known as phagocytosis. The paramecium, also a heterotroph, moves about by tiny hair-like projections known as cilia. Cilia also help the paramecium in engulfing food.

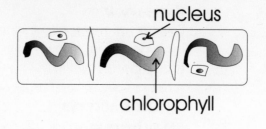

Algae

Eukaryotic algae

The plant-like protists, they include green algae, brown algae, golden algae, diatoms, dinoflagellates, red algae, and the euglena. The euglena, heterotrophic in the absence of light, possess chlorophyll and from one to three flagella.

Slime molds

Fungi-like protists that resemble fungi because at one point in their life cycle they come together as a unit—looking like a multi-celled organism—and form spores.

FUNGI

Formerly classed as plants, they do not possess chlorophyll—making them heterotrophs. The fungi are structurally different from plants so that they have been placed in their own kingdom. For example, they have no cell membrane. Some fungi are unicellular, while others are multi-cellular. Fungi are predominantly haploid, becoming diploid for the purpose of reproduction through the production of spores.

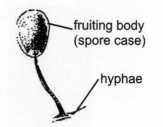

Reproductive body of Fungus

Types

Classed into three groups based on how they obtain nutrients.

A. Saprophytic
Live off dead organisms

B. Parasites
Live off live organisms

C. Symbiotic
Work in concert with other organisms

Members
The following are the major groups of fungi the AP Biology Exam will test on.

A. Molds
Molds invade hosts through root-like hyphae, which they also use for reproduction. Bread molds are an example of this group.

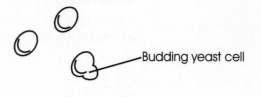

Yeast

B. Yeast
Along with mildews, they are primarily unicellular but form fruiting bodies containing ascospores for reproduction. This group also includes the much sought after truffle.

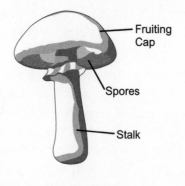

Mushroom

C. Mushroom
Multi-cellular fungi includes the well-known mushroom.

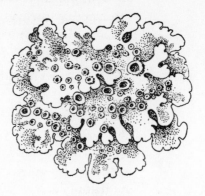

Lichen

D. Lichens

a symbiotic union between a fungi and either a chlorophyta [eukaryotic algae] or a cyanobacteria.

PLANTS

Multi-cellular and autotrophs, these organisms contain chlorophyll bound in organelles called chloroplasts—*chloro*phyll *plast*ids.

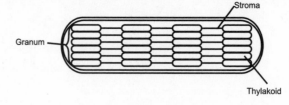

Chloroplast

The dominant generation in most plants is diploid, most have vascular tissue, and they reduce desiccation with a cuticle—a waxy coating on all aerial parts. Advanced forms possess ovaries and have adapted to seasonal changes as in the case of the deciduous trees. The student will want to use this outline as a guide to studying plants, which we start with Bryophytes despite the fact that the nonvascular Bryophytes are very different from the vascular plants.

Bryophytes

The simplest plants, these plants do not have true roots, stems, or leaves. This fact limits the movement of water through the organisms, and therefore is a major limitation to the plants, mostly in size, but also in storage of water and in the process of reproduction, so their habitat is very moist. In reproduction, for example, the sperm must swim to the egg. Adult bryophytes are haploid, entering the diploid stage when the sperm fertilizes the egg. Following this, spores are produced by meiosis, and alternation of generation has once again brought about the haploid generation, which dominates the cycle. This division of plants includes the mosses, liverworts, and hornworts.

Vascular plants

More typical of the plants are the vascular members of the Kingdom, some of which are extinct, but are included in the schema. Vascular plants are predominantly diploid throughout their life cycle, producing haploid sperm and egg for the purposes of reproduction. Eggs are produced in an ovary and sperm in the pollen grains.

A. Seedless vascular plants

1. *Lycophytes*—one group of lycophytes includes the extinct woody trees of the carboniferous period and the second includes epiphytic plants that live off other plants, such as the so-called club mosses because of their club-shaped, spore-producing cones.

2. *Sphenophytes*—includes another group of extinct trees from the carboniferous period and herbaceous plants known as horsetails because they produce filamentous leaves that resemble horsetails.

3. *Pterophytes*—this group includes the ferns and has spore-bearing capsules on the underside of their leaves, called sorl.

4. *Gymnosperms*—the first group of seed plants, some "naked seed" conifers, take up to three years to complete reproduction, with the result being an unprotected seed. Gymnosperms bear pollen-producing male cones and ovule-bearing female cones all on the same tree.

5. *Angiosperms*—include the flowering plants, a major evolutionary adaptation for the angiosperms and produce coated or covered seeds within an ovary on the flower.

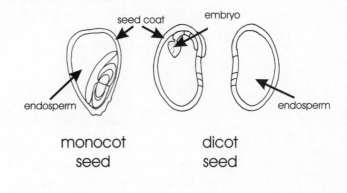

monocot seed   dicot seed

6. *Classification*

- *Monocots*—In seeded plants, the structure that stores the next generation's embryo, as well as a good amount of nutrients to get the embryo started, is contained in cotyledons. These are not seed halves, although the dicots contain two. Monocots only contain one.

- *Dicots*—so-called because they have two seed-bearing structures.

7. *Tissues*—there are three different types of plant tissues based on their function.

- *Ground tissue*—these tissues serve as support tissues for storage, photosynthesis, and secretion. Examples are collenchyme, sclerenchyme, and parenchyme tissues.

- *Dermal*—covers the outer portion of plants and include the guard cells, hair cells, stinging cells, and glandular tissue. The aerial tissues secrete the cuticle, a waxy coating that acts to prevent water loss.

- *Vascular*—for transporting material throughout the plant, which includes xylem and phloem. These tissues, respectively, carry water and minerals up the plant and distributes photosynthetic products throughout the plant. Xylem cells are perforated specialized cells producing, in effect, one long tube. These tissues play a vital role in the transpiration-tension theory that explains how through adhesive and cohesive forces—hydrogen bonding, surface tension, capillary action—and the loss of water through the stomata, a plant moves water, in some cases, over considerable distances. Phloem tissue is made up of sieve-tube cells and companion cells and is likewise "connected," helping to pass on nutrients.

8. *Organs*—vegetative organs are roots, stems and leaves. Reproductive organs are flowers, fruits, and seeds.

- *Seeds*—contain the next generations embryo as well as stored nutrients. Under the right conditions and with the absorption of water, the seed germinates. The seed then goes through a primary and a secondary growth on the way to maturing into the next generation of adult plant.

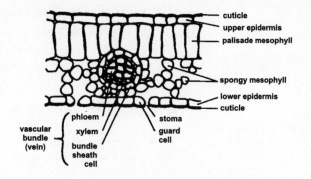

cuticle
upper epidermis
palisade mesophyll

spongy mesophyll
lower epidermis
cuticle

vascular bundle (vein)

phloem
xylem
bundle sheath cell

stoma
guard cell

Cross section of typical leaf

- *Leaves*—leaves are the main photosynthetic structure in plants and contain five distinct parts, each with their own function.

  ▷ *Epidermis*—the outer layer of cells, like the skin of an animal, it serves to protect and prevent desiccation. A waxy substance called a cuticle may be secreted.

  ▷ *Palisades layer*—so-called because it contains a layer of closely packed vertical photosynthetic cells. The arrangement allows the maximum exposure to the light of the sun. The cells are photosynthetic.

  ▷ *Spongy layer*—contains photosynthetic cells more loosely connected than those in the palisade layer. The spaces provide room for CO_2, O_2, and H_2O vapor to circulate.

  ▷ *Guard cells*—found on the underside of leaves, these are specialized epidermal cells that change shape as the amount of water in the leaf changes. The opening that results when they swell up with water is called the stomate. It provides for passage of gases in and out of the leaf, including water vapor.

  ▷ *Vascular bundles*—made up primarily of xylem and phloem tissues bundled together.

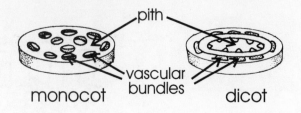

Typical angiosperm vascular bundles

- *Stems*—the main support of the plant, they also contain tissues for storage of materials as well as vascular tissue and, in some cases, cells with chloroplasts.

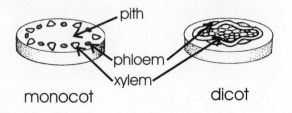

Cross section of typical angiosperm stem

- *Roots*—are the extensions of the stem, in a manner of speaking, into the ground and provide for water transportation, anchorage, some support, and, possibly storage.

9. *Flowers*—the reproductive organ of flowering plants, students should know the parts and their function.

Pistil

The female reproductive organ, it consists of a stigma, where pollen lands, and provides connection between the style and the ovary where the egg is fertilized. The ovary turns into the fruit containing the seed. The seed includes the embryo and all stored food for the embryo.

Stamen

The male reproductive organ, it consists of the anther and the stalk, or filament. The anther produces pollen that contains a generative cell and a tube cell. The pollen develops into a pollen tube, which grows down the style to the ovary. During the growth of the pollen tube, the generative cell develops into two sperm cells. In the ovary, one sperm fertilizes the egg, and the nucleus of the other sperm fuses with the polar nuclei.

Sepals

Sepals protect the unopened flower.

Petals

Petals in most cases aid in pollination by attracting pollinators.

1. *Hormones*—plants produce hormones that can pass from one cell to the other and affect the growth and differentiation of plant tissues.

   - *Auxins*—promotes the elongation of certain cells and helps in the growth of the plant, mainly in the tips of shoots and roots.

   - *Gibberellins*—these promote cell growth and fruit and seed development.

   - *Cytokinins*—as the name suggests, these stimulate cell division.

   - *Ethylene*—in the form of a gas, ethylene promotes the ripening of fruit and the production of flowers.

   - *Abscissic Acid*—a growth inhibitor that delays seed germination and bud development. It also may play a role in abscission.

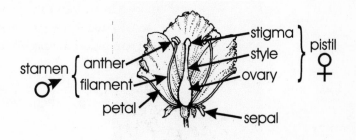

Flower

2. *Tropisms*—Plants are capable, under the influence of hormones, of growing toward or away from certain influences in the environments as they meet their needs in such an action. The name suggests a moving toward, but it is a growth process.

   - *Phototropisms*—growth toward light that is mediated by auxins.

   - *Gravitropism*—response by roots to the field of gravity and appears also to be mediated by auxins, although the classic role of light is absent—gravity being the influence that affects the concentration of auxins in certain cells.

   - *Thigmotropism*—response to touch.

3. *Photoperiodism*—the response, by plants, to varying amounts of light over time, particularly the length of day and night.

- *Long-day plants*—are stimulated to bloom by increasing amount of daylight or, conversely, it may also be a decreasing amount of dark.

- *Short-day plants*—are stimulated by decreasing amount of light to enter their growth stage—it may also be an increasing amount of dark.

- *Day-neutral plants*—do not respond to changes in daylight but to some other environmental factor.

ANIMALS

In this section, we will briefly outline the organisms that students should be familiar with. The schema used should be noted for the evidence it provides for evolution. The student should be familiar with these evolutionary connections. In an outline such as this, an AP Biology text is indispensable and provides the detail that would make this a text on zoology or comparative anatomy if it were included. AP Biology tests more on the related systems found in organisms, with comparisons to humans, than it does on the taxa of the individuals. Ancillary material to this subject would include the following: animals are multi-cellular heterotrophs, predominantly diploid; embryologically, most have three cell layers with increasing complexity as we go through the taxonomy; body symmetry is radial—a circular body plan—in the lower phyla or bilateral where organism possess a top (head), bottom (tail), front (ventral), and back (dorsal), and many organisms are segmented, which is a key evolutionary stage.

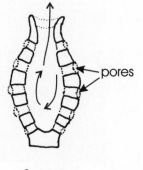

Sponge

Porifera

the sponges. While some look cigar-shaped, they have no symmetry and contain a loose association of spicules that form a support—not skeletal—system. As the name suggests, they contain pores through which water and food pass.

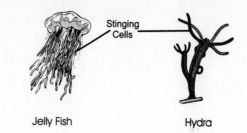

Jelly Fish   Hydra

Cnidaria

Cnidaria include the sea anemone, hydra, jellyfish, and corals, which secrete a hard coating around themselves for protection. There are two morphological forms, the medusa and the polyp. The medusa is free-floating. In some cnidaria, the two stages alternate throughout their life cycle, with the polyp from producing the motile medusae which produce the sperm and egg.

Planaria

Platyhelminthes

Platyhelminthes are planarians, flukes, and tapeworms. Planarians are free-living carnivores, flukes are either internally or externally parasitic and live off the fluids of the host, and tapeworms are internal parasites in the digestive tract of vertebrates. Tapeworms lack their own digestive system.

Roundworm

Aschelminthes

A. Nematodes

Roundworms are the first organism with a complete digestive tract, many are free-living and help recycle nutrients, and several are parasitic as is the trichina worm.

Rotifer

- *Rotifer*—microscopic with several specialized organs. They are primarily filter feeders.

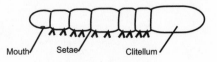

Earth Worm

Annelid

The segmented worms, they include the leeches, a blood-sucking parasite; the earthworms; and the polychaete worms, which occupy a mostly marine environment. At this point in the taxa, the organisms are becoming more and more complex by either possessing an entire organ system or by one or more organ systems attaining significant complexity.

Clams

Mollusk

Soft-bodied—although some possess a shell—this group includes the snail, bivalve, octopus, and squid.

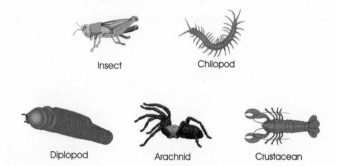

Insect   Chilopod

Diplopod   Arachnid   Crustacean

Arthropod

The jointed-legged animals, they possess exoskeletons of considerable importance and include the spiders, insects, crustaceans, the so-called "centipedes" or chilopods, and millipedes or diplopods.

Sea star

Echinoderms

Echinoderms include the sea stars, sea cucumbers, and sand dollars, all of which have radial symmetry, although larval body shapes are bilateral, which is believed to be an ancestral recapitulation.

Chordate and Mammal

Chordates

These animals, the highest in the taxa and on the evolutionary scale, all share the three following characteristics: (1) a flexible notochord (rod) at some time in their development that provides support for the nervous system and is located dorsally. The notochord is generally replaced by a vertebral column. (2) dorsal, hollow, nerve cord that in some differentiates into a brain and spinal cord. (3) pharyngeal gill slits that allow oxygen-carbon dioxide exchange. In the higher organisms the pharyngeal gill slits appear mainly as pharyngeal folds. There is also a muscular tail at some point in their development which some lose as early as the embryonic stage.

Subphylum—Vertebrate Classes

A. Pisces

B. Amphibian

C. Reptiles

D. Aves

E. Mammalian

MULTIPLE CHOICE QUESTIONS

1. The plant hormone responsible for fruit ripening is

   (A) abscissic acid

   (B) auxin

   (C) cytokinin

   (D) ethylene

   (E) gibberellin

2. Which of the following is found in all viruses?

   (A) protein coat

   (B) cell membrane

   (C) membrane-bound organelles

   (D) ER

   (E) mitochondrial DNA

3. Which of the following categories includes the most distantly related organisms?

   (A) family

   (B) species

   (C) class

   (D) genus

   (E) order

4. In flowering plants, sperm are produced by the

   (A) ovary

   (B) anther

   (C) megasporangium

   (D) sporophyte

   (E) generative nucleus

5. Which of the following is not a type of plant stem?

   (A) corm

   (B) node

   (C) rhizome

   (D) tendril

   (E) tuber

6. Gymnosperms are NOT

   (A) seed plants

   (B) predominantly diploid

   (C) flowering plants

   (D) conifers

   (E) naked seed plants

7. Viruses are

   (A) always viewed with an light microscope

   (B) are an enzyme-nucleus mix

   (C) obligate, intracellular parasites

   (D) cellular

   (E) host independent

8. An invertebrate is found in a freshwater setting, and studies show it to have three developmental body layers and a cuticle covering its outer body. It belongs in the same phylum as the

   (A) hydra

   (B) leeches

   (C) sea stars

   (D) sponges

   (E) crustaceans

9. Which of the following have a visceral mass and a muscular foot?

   (A) Medusoids

   (B) Annelida

   (C) Aschelminthes

   (D) Mollusca

   (E) Arthropoda

10. Which of the following is radially symmetrical and possesses nematocysts?

(A) Porifera

(B) Coelenterates

(C) Mollusca

(D) Amphibians

(E) Polychaetes

EXPLANATIONS AND ANSWERS FOR MULTIPLE CHOICE QUESTIONS

1. **The correct answer is (D).** Ethylene promotes the ripening of fruit and the production of flowers. Auxins promote the elongation of certain cells and helps in the growth of the plant, mainly in the tips of shoots and roots. Gibberellins promote cell growth and fruit development—not ripening—and seed development. Cytokinins stimulate cell division. Abscissic acid delays seed germination and bud development.

2. **The correct answer is (A).** Of the structures listed, all viruses are found within a protective protein coat. All of the other structures are found in eukaryotic cells: a cell membrane to act as a barrier to things entering the cell, ER, which refers to the endoplasmic reticulum, the canal system of the cell, and the site where protein synthesis takes place; membrane-bound organelles is self-explanatory, and mitochondrial DNA is found in the mitochondria.

3. **The correct answer is (C).** The most specific group listed is the species and includes organisms that are virtually identical to each other, with slight variations. They have the same structures, act alike, and can mate and produce offspring that can mate. This group has the most closely related organisms. The next group, less specific than the species, is the Genus. The genus *Canus*, for example, includes all the dog-like organisms—wolf, coyote, and the domesticated dog. The family level is one up from the Genus and lies several levels below class, which means it has more widely varied examples than the Genus level but far fewer than the class level. Class is the group, on the list, with the most widely varied organisms. For example, the class Mammalia of the phylum Chordates includes elephants, tigers, bears, cats, dolphins, and man.

4. **The correct answer is (C).** The anther is associated with the production of sperm in flowering plants but, more specifically, the structure more closely identified with sperm production in the anther is the generative nucleus. Sporophyte refers to the

dominant generation among the plant kingdom species and involves the entire organism. All other parts are associated with the female structures. The ovary is where the egg is produced and fertilized and will become the embryo; the megasporangium produces the mother spore cell in tracheophytes that leads to the production of the four haploid cells, one of which will become the egg; and the stigma is the place where pollen grains fall and ultimately grow a pollen tube to the ovary. It is supported by the stigma; both are female reproductive structures.

5. **The correct answer is (B).** A node is a site on the stem where leaves attach—and unattach—at the end of the growing season. They are *not*, therefore, an example of a stem. Of the other items in the question, all of which are a type of stem, tendrils assist climbing plants as attachment points and tubers; rhizomes and corms are all examples of underground stems. Tubers store large amounts of starch, as in the potato, rhizomes are found in ferns and are involved in vegetative propagation, and corms are specialized leaves that can store food.

6. **The correct answer is (C).** Gymnosperms have what are often referred to as "naked seeds," but they are seeds nonetheless. Like all tracheophytes, they are predominantly diploid and they bear—both the smaller male and the larger female—cones. However, a gymnosperm is not an angiosperm or a flowering plant.

7. **The correct answer is (C).** In order to reproduce, viruses need cells that they can invade. These two facts make them obligate and parasites. In addition, they are found intracellularly, taking over the cell only for the purposes of reproduction. In order to see viruses, they must be viewed with an electron microscope since they are smaller than bacteria. It is also well known that they contain a protein coat and a piece of DNA, not enzymes and certainly not a nucleus. They are also not cellular entities and since they are not free living, they are definitely host dependent. Host dependent can mean they need a host to survive, or it can mean they need a host to reproduce. In the case of this question and the choices given, the latter is taken to mean host dependent here.

8. **The correct answer is (E).** The organism in question cannot be an echinoderm, as they are exclusively marine, so it is not a sea star. Sponges and coelenterates do not have a middle body layer and annelids do not have an outer cuticle. Crustaceans match the description given in the question.

9. **The correct answer is (D).** Mollusks include the bi-valves, which possess considerable visceral mass, and a muscular foot

with which they burrow and use for movement. Medusoids include the very simple body planned coelenterates, such as hydra and jellyfish, and lack significant visceral mass—if what they possess can be called that. Annelids do not have a muscular foot, and their viscera, while well developed, is not as substantial as a bi-valves. Aschelminthes is a roundworm with more viscera than the cnidarians, but still considerably less than even the annelids. Finally, the arthropods are jointed-legged creatures and do not possess a single, muscular foot.

10. **The correct answer is (B).** Coelenterates, such as the jellyfish and the sea anemone, are well known for their radial symmetry and their stinging cells. Porifera have no symmetry. All the rest of the choices have bilateral symmetry—polychaetes are segmented worms that are predominantly marine; molluscs include octopuses and squid, neither of which have stinging cells; and amphibians are chordates with neither radial symmetry nor stinging cells. Some frogs secrete powerful toxins through their skin, but none of the cells in their skin has the ability to sting.

Vocabulary

abdomen

abscissic acid

aerobes

algae

alternation of generations

amimalia

amoeba

angiosperm

animal

annelid

anterior

anther

apical

arachnida

auxins

bacillus

bacteria

basidium

bilateral

binomial nomenclature

bread mold

bryophyte

budding

budding

bulbs

cambium

cephalization

chilopoda

chitin

chordata

cilia

class

cnidaria

coccus

coelenterate

coelom

crustacean

cyanobacteria

cytokinins

day-neutral

dicots

diploid

diplopoda

dorsal

echinoderm

ectoderm

egg nucleus

endoderm

endosperm

ethylene

euglena

euglena

facultative anaerobes

family

filament

flagellum

fruiting body

fungi

gametophyte

genus

gibberellins

grafting

gravitropism

gymnosperm

haploid

hyphae

insecta

kingdom

legumes

lenticels

lichen

Linnaeus

liverworts

long-day plant

mantle

meristem

mesoderm

mildew

mollusca

monera

monocots

mosses

mycelium

mycorrhizae

N2-fixing bacteria

nematode

obligate anaerobes
order
ovary
ovule
paramecium
parasite
parasites
penicillin
petals
phagocytosis
phloem
photoperiodism
phototropism
phylogeny
phylum
pilli
pistil
plantae
platyhelminthes
polar nuclei
pollen
porifera
posterior
protista
radial
roots
rotifer
runners
saprophytes
segmentation

sepals
short-day plant
sieve tubes
species
sperm nucleus
spirillus
spores
spores
sporophyte
stamen
stigma
style
symbionts
symmetry
taxonomy
thigmotropism
thorax
tracheids
tracheophyte
tropisms
truffle
tubers
vascular
ventral
viruses
xylem
yeast

Chapter 7
ANIMALS—STRUCTURE AND FUNCTION

OVERVIEW

Approximately 32 percent of the AP Biology exam is on plant and animal structure and function. As was mentioned in Chapter 6, all of the material relates to all of the other material. To this end, we point out that some of this material may be repetitious, never a detriment to learning, however. In this chapter, we will outline the animal systems, focusing on human systems. These are the areas students can expect to be tested on the most, making it essential that much detailed study beyond this outline be done. A very broad area not discussed in this outline is that of hormones in heterotrophs other than humans. Essentially, these hormones fall into three categories: sex attractants, territorial markers, and chemical facilitation of internal control, the latter being much too broad to discuss here. However, the student should be familiar with the area dealing with phero-mones. A word about cell differentiation—or specialization—in multicellular organisms is appropriate here: the good news is that cells are specialized to do what they do very well. The bad news is that all those specialized cells have become interdependent, thus the discovery of increasingly complex systems in increasingly complex organisms. Students should also be familiar with the sequence from cell—tissue—organ—system—organism and homeostasis.

DIGESTIVE SYSTEM

HUMANS

Large nutrient molecules are not usable by heterotrophs; they must be broken down into smaller molecules that are capable of being used by cells. This is the part of digestion called catabolism and includes two main types: mechanical and chemical. For the most part, mechanical starts the process in most organisms, with chemical becoming increasingly prevalent as the process proceeds. Digestion—more specifically catabolism—replaces the water molecule in a process called hydrolysis that is the reverse of dehydration synthe-sis—an anabolic reaction—where water is removed from molecules as polymers are formed. Digestion also includes the making of smaller pieces of nutrients from larger pieces to facilitate absorption, as in the case of fat molecules. The second process discussed here, of course, is absorption of the catabolized nutrients. In humans, mechanical and chemical digestion both start in the mouth or oral cavity. Chewing is aided by the teeth, the tongue, and even the cheeks and lips as they help position the food, as does the tongue. The teeth help to tear, shred, and grind the nutrients, which increases the surface area. For example, the surface area of a cube that is 1cm on a side is 6cm^2. If we cut it into 1,000 pieces—each having a .1cm on a side—the surface increases tenfold to 60cm^2. Thus, mechanical digestion clearly aids chemical digestion.

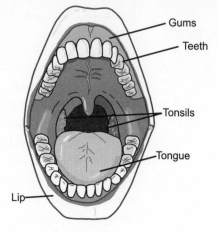

Mouth

In the mouth, the carbohydrates are acted on with amylase, an enzyme secreted by the salivary glands, that breaks the starch into smaller polysaccharides. The pharynx is a connection between the mouth and the esophagus. It houses, among other things, the flap of skin called the epiglottis that should close over the larynx and trachea each time we swallow. Swallowed items will get tangled in this area if one is breathing and swallowing at the same time. The connection between the mouth and the stomach is the esophagus. Nutrients are passed from the mouth to the stomach by a series of alternate muscle contractions along the esophagus. One set of smooth muscles ring the length of the esophagus and another set run vertically along its length. Each set of muscles has the capability of alternately contracting. The effect is an opening and closing of a section of the esophagus as food passes along its length. Contrary to popular belief, one can drink water upside down and even in space.

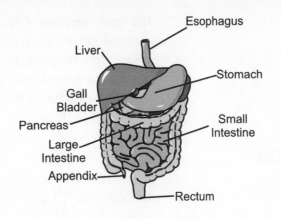

Digestive System

At the entrance to the stomach, we find a sphincter muscle that closes to keep materials in the stomach and to very briefly allow for the slow passing of materials into the stomach. As it is positioned anatomically so near the heart, it is called the cardiac sphincter. Irritation and illness sometimes cause this sphincter to allow for the reverse movement of material. In the stomach, proteins are acted upon for the first time, and physical breakdown continues. Pepsin, an enzyme activated by the acid environment in the stomach, acts on peptide bonds to begin the breakdown of proteins. Both substances are secreted under the influence of hormones as well as by nerve impulses. The lining of the stomach is protected by a balanced secretion of mucus. The partially digested material, now known as chyme, is gradually allowed to exit the stomach through the pyloric sphincter in small amounts for further digestion in the upper portion of the small intestine, where digestive chemicals from the gall bladder and the pancreas are secreted. As chyme leaves the stomach, it is, of course, very acidic. As a digestive organ, the pancreas secretes no less than four digestive enzymes that act on proteins and lipids: trypsin, chymotrypsin, pancreatic amylase, and pancreatic lipase, all packaged in an alkaline solution to neutralize the acid from the stomach. The digestive enzymes act to continue the digestion of sugars and proteins. The pancreatic lipases are aided by bile in acting on lipids. The liver produces a substance called bile—manufactured from recycled RBCs—which helps to emulsify fat molecules to promote absorption of certain fat molecules into the lymph system in the villi of the small intestine. Beyond the one section immediately distal to the stomach, the small intestine is divided into two distinct sections: digestion just described will occur in the upper section and absorption of catabolized nutrients will occur in the lower section through structures called villi that resemble finger-like structures that project inward along the length of the lower small intestine. The upper section continues the digestion of all three major classes of nutrients.

The large intestine is the next part of the digestive system and is involved mostly in absorption of water. Species of bacteria also reside in the large intestine that help manufacture vitamin K, a component necessary for the clotting of blood. The remaining material, called feces, exits the body through the anus, after it is stored in the rectum for a short time.

COMPARATIVE

Most coelenterates subdue their prey with stinging cells and digest through intracellular and extracellular digestion. As they burrow through the soil, earthworms grind food mixed with soil in their gizzards and pass it on to be absorbed in the intestines; the undigested material then passes out of the body. Although not the same structurally as earthworms, insects also possess a crop and a gizzard. Gastric glands secrete digestive enzymes; absorption takes place mainly in the stomach, after which undigested material is passed to the intestine and exits through the rectum.

RESPIRATORY SYSTEM

HUMANS

Cellular respiration, a key energy-releasing reaction in living organisms, requires oxygen and produces carbon dioxide. Carbon dioxide is the gas that, through the production of HCO_3, regulates breathing. These two gases must be exchanged by way of the respiratory system and the circulatory system. Air is filtered, warmed, and moistened as it enters the nasal passage through the nose and then passes through the pharynx on its way to the trachea. Air passes through the vocal cords, which are housed in the larynx, on its way through the trachea to the lungs. Ringed with C-shaped cartilage, the trachea has structural integrity while at the same time being flexible. If it were bony material, almost any blow that would strike this area could be disastrous.

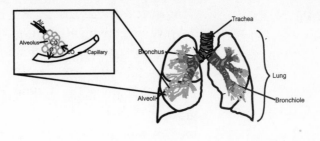

Lungs

The trachea branches into two tubes called the bronchi, each of which passes into the lungs through even smaller and more numerous tubes called bronchioles. The main area for gaseous exchange is the lung. Each lung consists of several lobes or sections. The chest muscles, both inner and outer along with the diaphragm, create volume changes in the lungs, which establish pressure gradients between the outside air and the lung cavity.

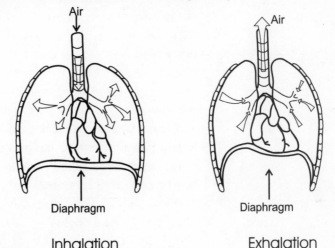

The changes in pressure alternately force air in and out of the lungs. Small sacs called alveoli form clusters like grapes and are surrounded by capillaries, where gas exchange actually takes place. Arriving at the alveoli, the blood contains large amounts of carbohemoglobin. This facilitates the diffusion of carbon dioxide into the alveoli through a single cell layer separating the open area of the air sacs and the capillaries. Housed in the air sacs is a large quantity of oxygen that diffuses out of the air sacs into the capillaries. The oxygen meets up with hemoglobin that recently dumped its carbon dioxide to form oxyhemoglobin, which is then transported to the cells that carry on cellular respiration. The reverse of this process occurs in the cells.

Protists achieve the proper gas exchange through simple diffusion of gases with their immediate environment. Coelenterates achieve gas exchange by diffusion through two cell layers. Annelids exchange gases through a moist skin. Large numbers of the members of this group flee their burrows during large amounts of rainfall since the oxygen supply in the water that enters their burrows is soon depleted. Arthropods flex their abdomen, causing oxygen and carbon dioxide to be exchanged through openings called spiracles.

CIRCULATORY SYSTEM

HUMANS

Many substances need to be moved around the body, and this system achieves it through the use of muscles. Humans possess a closed circulatory system where the fluid for transport is kept within transport vessels. Some leakage of the fluid portion of the blood does occur, but it is picked up by the lymph system and returned to the closed system. (Note: Leakage is a normal function that allows nutrients to leave the capillaries and arrive at the cells.) In the veins, we find valves every few centimeters that point in the direction of the heart, so that when they close, blood is prevented from back-flowing away from the heart. Muscle contractions in the skeletal muscles push the blood forward, helping the blood to return to the heart in the venous system.

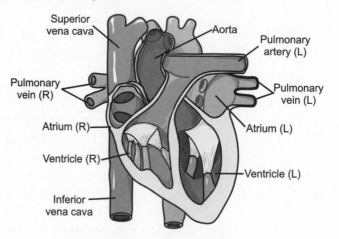

Heart and associated vessels

Blood is transported away—*Arteries Away*—from the heart by the action of the cardiac muscle in vessels called arteries. Blood is returned to the heart with the aid of skeletal muscles in vessels called veins and a series of valves in the veins. The veins connect through the smallest vessels of all called capillaries, the sequence being heart—arteries—arteriole—capillaries—venule—veins—heart. The heart, the main pumping organ in the circulatory system, is actually two pumps in one. More accurately, one half beats in concert with the other half, delivering blood to two separate areas of the body. The right side of the heart receives low-oxygen blood in the upper part, known as the atrium, and uses the lower part to pump it through pulmonary arteries to the lungs in what is known as pulmonary circulation. The left side of the heart receives high-oxygen blood in the upper part, known as the atrium, and uses the lower part to pump it first through the aorta—the largest artery in the body—and then through arteries to the body.

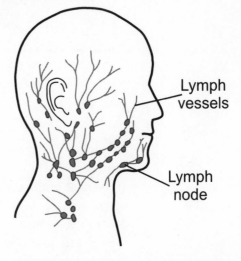

A portion of the lymph systems

Blood is a fluid that circulates oxygen, carbon dioxide, nutrients, waste materials, hormones, vitamins, and anything else that the body needs to transport and some things it does not need to transport. It is composed of about 55 percent liquid, known as plasma; the rest consists of solid parts and includes erythrocytes, leukocytes, and platelets. Blood types were discussed in an earlier chapter.

COMPARATIVE

Protists "swim" in their environment, which includes nutrients and oxygen. This allows anything to diffuse in or out quite easily. Coelenterates, like the protists, exchange materials with their environment simply by diffusion. Annelids possess closed circulatory systems that contain a simple muscular loop.

IMMUNE SYSTEM

HUMANS

The basic functions of the human immune system include recognition of an invading entity, as well as abnormal body cells, and the attempt to eliminate it. Lines of defense against attack include the skin, anti-microbial proteins, competing bacteria, cilia, and gastric juices.

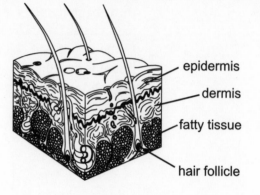

Cross section of Human skin

In addition, humans have developed means of augmenting the immune system through the use of vaccines, antibiotics, or the passing of antibodies from one person to the next. In addition, interferon has been identified as help in the invasion of viruses. Macrophages and neutrophils engulf and digest invaders while the B-cells and T-cells work in concert in both the interstitial areas, as well as within cells, to eliminate the antigens. The organs involved in human immune responses include the bone marrow and the thymus, where interaction between B-cells and T-cells is mediated; lymph nodes; the spleen; adenoids; and tonsils. Tissues in the appendix may also have a role similar to the tonsils.

COMPARATIVE

Relatively little emphasis is placed in the AP exam on immunity outside the study of humans and their immune system. Invertebrates, for the most part, form alliances with competitive microorganisms or possess physical barriers in response to attack by foreign bodies. In most instances, they flourish where there are no attackers and cease to exist where there are attackers, reaching a level of homeostasis with the environment. Humans, on the other hand, carry their immune system with them or modify it with augmentation.

EXCRETORY SYSTEM

Waste must not be allowed to accumulate in or around cells and so is gathered and eliminated by an excretory system. Homeostasis, maintaining a steady state in humans, is a critical role of this system.

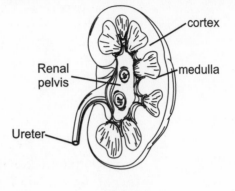

Kidney

The metabolism of proteins, for example, produces toxic substances harmful to body tissues and, hence, must be gotten rid of. It is important to distinguish between the elimination that is achieved in the digestive system and the excretion done in this system. Material that passes entirely through the system is finally eliminated from the body as a passive act. What ends up in the lower large intestine is material that is left over after all the other nutrients have been acted on. It is not gathered and deposited there as an active process of excretion performed on materials, such as proteins and carbohydrates; it just happens to end up there. On the other hand, excretion is a homeostatic event with several systems working in concert to actively gather the amount necessary to maintain homeostasis and excrete those harmful substances from the body. The metabolism of proteins produces nitrogen wastes, most predominantly ammonia. This is coupled with carbon dioxide in the liver and "neutralized" or made less harmful in the form of urea.

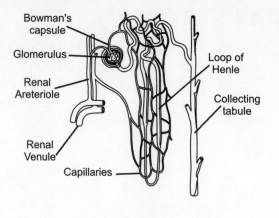

Nephron

Transported to the kidneys, blood is filtered out in units called nephrons that house the main filtering structure, Bowman's capsule and tubules. Mediated by hormones and other chemical factors, the resulting mix that has been selectively separated is passed to the bladder through the ureters, where it is stored until it will be excreted from the body through the urethra.

COMPARATIVE

Unicellular organisms simply pass their nitrogenous wastes through their plasma membrane by diffusion and maintain water balance by osmosis. Annelids filter and collect these wastes in nephridia that connect to the outside environment. Arthropods collect nitrogenous wastes in structures called Malpighian tubules and then pass the wastes to the environment. Aquatic organisms, for the most part, excrete ammonia directly into the water, and birds produce uric acid, a largely insoluble substance that precipitates in their systems, which they then excrete.

NERVOUS SYSTEM (NS)

HUMAN

The basic unit of the nervous system is the neuron. Three principal neurons are the sensory (afferent), motor (efferent), and interneuron (association), which, respectively, receive signals, activate muscles, and connect the two and make up the neurons found in the spinal cord and brain.

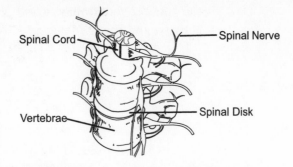

Segment of Human spinal column

Nerve cell axons are covered by a protective sheath called the myelin sheath, which is also where the segments of sheath connect exposed areas known as the Nodes of Ranvier's. Communication within the nervous system proceeds by way of an electrochemical event called a nerve impulse. As the nerve impulse proceeds along the nerve, an action potential is created by the interaction between ions inside and outside, the result of which is an electrical impulse. When the nerve impulse reaches the gap between two nerve cells called the synapse, a neurotransmitter manufactured in sacs or vesicles at the end of a one-nerve ending is dumped into the synapse, which causes the next nerve to "fire." One of the neurotransmitters, acetylcholine, is then broken down almost immediately—lest the next nerve keep firing without an impulse from the preceding nerve—by an enzyme called acetycholinesterase.

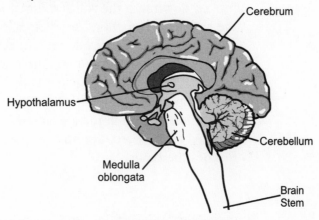

Cross section of Human brain

The central nervous system (CNS) is composed of the brain and the spinal cord. The main parts of the brain are the cerebrum (the largest) which controls all voluntary action and interprets sensory information; the cerebellum, or "little cerebrum," which coordinates muscle activity; the hypothalamus, which regulates homeostasis and secretes hormones; the pituitary, part of the hypothalamus, which is the "master" controlling gland of the body that controls hormone activity; and the medulla, which controls involuntary actions.

Peripheral nervous system

The peripheral nervous system (PNS) connects the CNS with all the outlying organs from the brain, skin, and all blood vessels. It is broken down into the somatic functions or voluntary functions and the autonomic functions or involuntary ones. The autonomic nervous system is further broken down into the sympathetic and parasympathetic portions of this system. The former prepares the body for action and the latter brings the body back to homeostasis. Finally, a rapid response, which initially goes from sensory neuron to an association neuron in the spinal cord and directly to a motor neuron, is called a reflex arc. Eventually—milliseconds later—the information reaches the brain.

COMPARATIVE

Single-celled organisms act mostly by moving away from irritants and toward attractants. Food attracts them, and toxins irritate them. Some even move according to the amount of light, but they possess no nerve structures. Coelenterates have a simple internal nerve net that covers their body. Annelids have a rudimentary nervous system consisting of a group of nerve cells called a ganglion and a nerve net that is the length of their body. This makes them capable of sensing their environment, but on a very primitive level, much like the protists. Arthropods, similarly, have a rudimentary system that is different in some structures from the annelids but similar in the resulting action. They have several specialized organs, such as eyes.

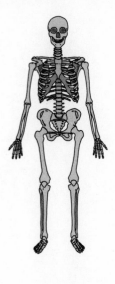

Skeleton

SKELETAL SYSTEM

HUMANS

Humans have an endoskeleton made of hard tissue called bones. Bones act to support, protect, make blood cells, store minerals, and aid—along with the muscle system—in movement. Students should be familiar with the process of ossification, Haversian canals, and the spongy and hard bony areas of a typical bone.

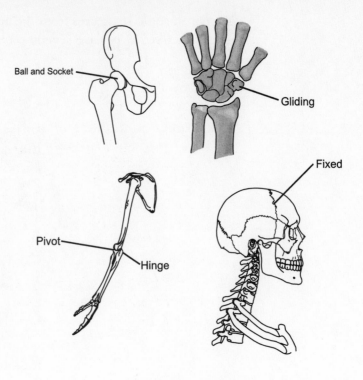

Human joints

The place where bones meet is called a joint, and humans have joints described by names that suggest their action, such as pivot, gliding, saddle, hinge, ball-and-socket, and even fixed joints.

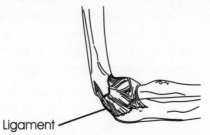

Bones are held in place by ligaments that in concert with muscles, help maintain the structural integrity of the body under normal conditions.

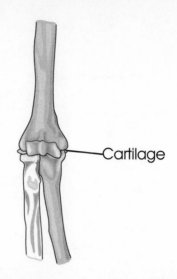

In between many joints is a smooth, cushioning material called cartilage; some structures are composed of cartilage, such as the nose and the outer ear.

COMPARATIVE

Protists, coelenterates, and annelids have no skeletal structures but possess a hydrostatic skeleton (support comes from an internal water pressure), and arthropods have an exoskeleton made of chitin.

MUSCULAR SYSTEM

HUMAN

Humans have muscle tissue that is capable of contracting and relaxing and that is positioned around the body.

Muscle system

The primary function of this tissue is to help in movement. Striated muscle, known also as skeletal or voluntary muscle, aids in moving the body along with the help of the skeleton—thus the name skeletal. Skeletal muscles are capable of very fast action, although they fatigue over a relatively short time. They are attached to the skeleton in opposing pairs: when one contracts, the opposite "partner" relaxes and vice versa. Humans also have a muscle type, known as smooth muscles, that comprise internal organs and blood vessels capable of much slower action but rarely fatigue. The third type of muscle that humans have is exclusively in the heart and is therefore called cardiac muscle. In cardiac muscle, we find a combination of striated and smooth muscle tissue all in one. The best of both of these two worlds then is achieved in cardiac muscle, which acts fast and does not fatigue under normal conditions. Students will want to know the details of muscle contraction and the link with the nervous system.

COMPARATIVE

Protists do not have muscles and are capable of changing the shape of the plasma membrane to achieve a variety of movements. Co-elenterates have a simple network of tissue that can contract. Annelids have simple muscle tissue that can provide them with more directed movement—one set of muscle tissue rings their body, and another set runs the length of it. Arthropods, with their exoskeleton, have muscles internal to that skeleton, which means that those muscles work in an opposite fashion to human muscles, as the "lever" system is opposite to that of humans.

ENDOCRINE SYSTEM

HUMAN

The other system in humans that acts to control the internal environment is the endocrine system. Composed of a series of ductless glands—they use the blood vessels as their tubes—these glands secrete a class of chemicals known as hormones. These chemicals have specific effects on specific tissues, known as target tissues, in the body. Once again, as in the nervous system, the basic need is to maintain a steady state or homeostasis. The so-called master gland that releases many hormones that affect other endocrine glands as well as significant areas of the body—bones and muscles in mediating growth—is the pituitary gland. The anterior pituitary secretes no less than six hormones that control growth (GH), the adrenal cortex (ACTH), the thyroid (TSH), the ovaries (FSH and LH), and production of milk (Prolactin). The posterior pituitary stores another two hormones, called vasopressin, that control the balance of water in the body and oxytocin, a key hormone in parturition. A list of the glands and what they control is offered in the following chart.

Gland →	Secretes →	Which Stimulate
Pituitary	TSH	Thyroid gland
anterior	ACTH FSH LH PRL GH	Kidney to secrete glucocorticoids Oogenesis and Spermatogenesis Oogenesis and Spermatogenesis Production of Milk Growth
posterior	ADG Oxytacin	Reabsorption of water Release of Milk
Pancreas	Glucagon Insulin	Increase blood sugar Decrease blood sugar
Adrenal Gland	Epinephrine	Increases blood sugar, constricts blood vessels
medulla	norepinephrine	Increases blood sugar, constricts blood vessels
cortex	glucocorticoids mineral corticoids	Increases blood sugar Reabsorption of Na^+ and excretion of Na^+
Thyroid	Thryoxin Calcitonin	Increases cellular metabolism Decreases blood Ca^+
Parathyroid	PTH	Increases blood Ca^+
Human glands and their secretions		

REPRODUCTIVE SYSTEM

HUMAN

The system given to the process of continuing the species is, of course, the reproductive system, and students should be conversant with where the events take place in both male and female, when they take place, and under the influence of what hormones.

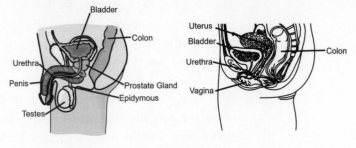

Human reproductive systems
(side view)

The AP exam will test on both the male and the female parts and their function. In addition, students are tested on critical events, such as gastrulation, blastulation, and general morphogenesis of the embryo to fetus.

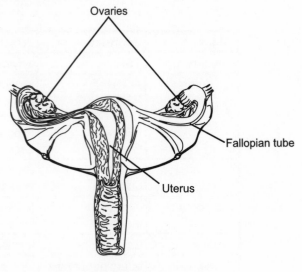

female reproductive system
(front view)

Females manufacture an ova in a structure called the ovaries. The cyclic secretions of hormones normally causes the release of one egg approximately every month. Generally, the process, called menses, is started with the production of an egg under the influence of FSH. The follicle, in turn, produces estrogen that triggers an increase in LH from the pituitary, turning the follicle into a body known as a corpus luteum, which continues to produce estrogen and a new hormone called progesterone. Progesterone readies the body for pregnancy by mainly increasing the number of blood vessels in the uterus. As the follicle ruptures under the influence of LH, it releases the egg in a process called ovulation. If the egg is not fertilized, the corpus luteum eventually shuts down, and the unfertilized egg, along with the now unnecessary, newly manufactured blood tissues in the uterus, are sloughed off and exit the body. This signals menstruation, and the process begins all over again as the level of FSH rises—there being no hormone such as estrogen to supress—in the other ovary. Thus the cycle repeats.

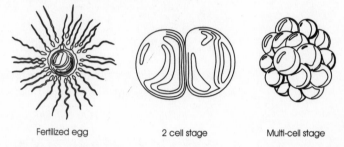

Fertilized egg   2 cell stage   Multi-cell stage

If the egg is fertilized, the embryo itself produces a hormone that continues the LH hormone to be secreted and suppresses FSH until birth. The fertilized egg divides repeatedly until it forms a ball of cells called the blastula. Eventually, this ball enters gastrulation as it differentiates into three cell layers forming the gastrula. Development of the notochord, found in chordates, and the neural tube means the embryo has entered the neural stage or the setting down of the precursor of the nervous system.

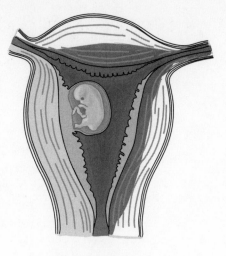

Embryo

At the age of three months, it is referred to as the fetus until the birth process or parturition is initiated.

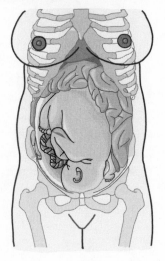

Fetus

Human males produce sperm in a process described in Chapter 4 on meiosis. This process is not cyclic but continual and is initiated and maintained by the hormone testosterone.

COMPARATIVE

Heterotrophs can reproduce asexually, which several phyla are capable of, but most can reproduce sexually, too.

MULTIPLE CHOICE QUESTIONS

1. Which of the following is the specialized absorption structure in the intestine?

   (A) alveoli

   (B) villi

   (C) Bowman's capsule

   (D) salivary glands

   (E) pyloric sphincter

2. Which of the following tissues contains the highest number of mitochondria?

   (A) nervous

   (B) skin

   (C) connective

   (D) muscle

   (E) bone

3. Select the correct sequence for human circulation.

   (A) heart—vein—capillary—artery—heart

   (B) heart—artery—lung—vein—body systems—heart

   (C) heart—artery—lungs—vein—heart—body systems

   (D) body system—lungs—heart—vein—capillary—artery

   (E) heart—artery—lungs—body systems—vein—heart

4. The skeletal system does all of the following except

   (A) support

   (B) transport

   (C) protect

   (D) digest

   (E) store

5. Endocrine glands

   (A) secrete vitamins

   (B) cease functioning after adolescence

   (C) begin functioning in adolescence

   (D) have ducts

   (E) have no ducts

6. Bile

    (A) activates pancreatic lipases

    (B) emulsifies fat droplets

    (C) congeals fat droplets

    (D) digests fats

    (E) is produced in the pancreas

7. Which of the following enzymes acts on protein?

    (A) pepsin

    (B) ptyalin

    (C) amylase

    (D) maltase

    (E) sucrase

8. Digestion in protozoans is

    (A) extracellular

    (B) intracellular

    (C) intercellular

    (D) dorsal

    (E) none of the above

9. Which of the following is true about the blood?

    (A) Arteries carry oxygen-rich blood.

    (B) Arteries carry oxygen-poor blood.

    (C) Veins carry oxygen-rich blood.

    (D) Veins carry oxygen-poor blood.

    (E) None of the above is true.

10. Chemical digestion begins in the

    (A) mouth

    (B) esophagus

    (C) stomach

    (D) gall bladder

    (E) small intestine

EXPLANATION OF ANSWERS FOR MULTIPLE CHOICE QUESTIONS

1. **The correct answer if (B).** The alveoli are structures that allow the passage, not absorption, of carbon dioxide and oxygen into the lungs. Bowman's capsule is where the filtration of the blood occurs in the nephrons of the kidney. Salivary glands produce

Peterson's AP Success: Biology

amylase that helps begin the breakdown of carbohydrates in the mouth, and the pyloric sphincter is the valve-like structure at the end of the stomach that allows the passage of chyme. The villi are specialized, finger-shaped structures in the lower intestine that are designed for absorption of digested nutrients.

2. **The correct answer is (D).** One would expect that the cells with the highest potential for activity would have the highest number of mitochondria, the powerhouse of the cell, which would be muscles, choice (D). Connective tissue has very little need for such a high level of activity, and bone cells reach the point of utility as fairly sanguine cells. The only choice that comes close to muscles would be nerve cells, but, while they may achieve a high level of activity from time to time, they certainly do not need to fire nearly as often as muscle cells. In addition, muscle cells need the energy when they contract and then to reconstruct the actin and myosin complexes when they relax to be ready to function again.

3. **The correct answer is (C).** While presented with a veritable hodgepodge of choices here—which just *may* occur on the exam—the only one that makes sense is choice (C), where the blood leaves the heart through an artery—*A*rteries *A*way—to the lungs, back to the heart through a vein, and from the heart, out to the body systems. Choice (A) looks good on the surface, except a vein is leaving the heart, not an artery. In choice (B), once the blood goes to the lungs, it goes right to the body systems, which erroneously bypasses the heart. Choice (D) also bypasses the heart as the blood returns from the body systems, and choice (E), like choice (B), bypasses the heart.

4. **The correct answer is (D).** The skeletal system does not digest anything for the body. It stores minerals, such as calcium and phosphorous; it supports the body mass and also helps with transport of the body in concert with the muscles; and its cells reproduce and make more bone cells.

5. **The corrrect answer is (E).** It is important to define endocrine glands as ductless—they offer a separate delivery system from gland to target area. The endocrine glands use the "ducts" of the circulatory system, also known as blood vessels. The exocrine glands, such as the salivary glands and the pancreas, have ducts. The endocrine glands begin functioning at parturition and continue to function throughout life. Finally, endocrine glands do not secrete vitamins.

6. **The correct answer is (B).** Bile is produced in the liver, and its main function is to emulsify fat droplets into smaller fat droplets, contrary to congealing—which it does NOT do. Congealing is the

gathering of small droplets into bigger ones. Pancreatic lipases come in an activated form, and bile certainly does not digest fats; it emulsifies them.

7. **The correct choice is (A).** Choices (B) through (E) represent enzymes that act on carbohydrates, ptyalin being another name for amylase. Maltase and sucrase, as their names suggest, are enzymes that act, respectively, on maltose and sucrose. Choice (A) is a protease that arises from a precursor, pepsinogen, that is activated by hydrochloric acid, which is the acid found in gastric juices.

8. **The correct answer is (B).** Unicellular protists take their food into the cell that is the entire organism and digest it entirely intracellularly. Humans, for example, digest their nutrients extracellularly and absorb them into cells only after they are digested. Since protists are unicellular, the suggestion of *inter*cellular digestion is meaningless. The reference to dorsal is a certain orientation or place on multicellular organisms.

9. **The correct answer is (E).** A favorite trap of the test-makers is to determine how solid the students are in their knowledge of the circulatory system by presenting questions that suggest that the movement of blood through the various vessels is wholesale. For example, they may suggest that arteries always carry oxygenated blood. While the bulk of blood carried in arteries is oxygenated, the blood in the pulmonary arteries is oxygen-poor—thus the reason for the trip to the lungs. Likewise, one can make a confusing statement about veins, but it would also be false. Veins that carry the blood back to the heart from the lungs are the only ones that carry oxygen-rich blood.

10. **The correct answer is (A).** Chemical digestion in humans can begin as soon as a nutrient enters the mouth, of course, if it has carbohydrate in it. So no matter what the nutrient is, it COULD begin in the mouth. The stomach is the second place that chemical digestion occurs when proteins are acted upon by pepsin. The esophagus is a tube that connects the pharynx with the stomach, so no digestion occurs there. No digestion takes place in the gall bladder—it is just a storage place for bile—and the small intestine is too general an answer. Digestion takes place in the upper portion of the small intestine and absorption takes placer in the lower portion. It doesn't matter, actually, since the question asks where digestion *begins*.

VOCABULARY

acetylcholine

acetylcholinesterase

ACTH

action potential

adrenal cortex

adrenal medulla

AIDS

alveolus

ammonia

amnion

antibody

antigen

anus

aorta

aortic valve

arterioles

artery

autonomic nervous system

axons

B-lymphocytes

bile

bladder

blastula

bolus

bone

Bowman's capsule

bronchi

bronchioles

capillaries

capillary

cardiac muscle

cartilage

cell body

cerebellum

cerebrum

chorion

chyme

cleavage

CNS

collagen

collecting duct

corpus luteum

dendrites

deoxygenated

dermis

digestion

emulsify

endocrine

epidermis

epiglottis

epinephrine

erythrocytes

esophagus

estrogen

expiration

fallopian tube

feces

feedback

fertilization

fetus

filtrate

FSH

gastrula

glomerulus

glucagon

glycogen

growth hormone

homeostasis

hormones

hyperthyroidism

hypothalamus

hypothyroidism

immune

inferior vena cava

inspiration

insulin

integument

interneurons

involuntary muscles

islet of Langerhans

joints

kidney

large intestine

larynx

left atrium

left ventricle

leukocytes
LH
ligaments
loop of henle
lungs
lymph
lymphocytes
marrow
mastication
medulla
menstruation
morphogenesis
motor neurons
mouth
myelin
nephridia
nephrons
neurons
neurotransmitter
notochord
ova
ovaries
oviduct
ovulation
oxygenated
pancreas
parathyroids
pepsin
peristalsis
pharynx
pituitary
plasma
platelets
PNS
polarized
progesterone
puberty
pulmonary
pyloris
reabsorb
rectum
renal

right atrium
right ventricle
saliva
Schwann cells
secretion
sensory neurons
sinuses
skin
small intestine
sodium-potassium pump
sperm
steroid
striated
subcutaneous
superior vena cava
sympathetic nervous system
synapse
systemic circulation
T-helper cells
T-killer cells
T-lymphocytes
tendons
testosterone
thyroid
thyroxin
trachea
trypsin
urea
ureters
urethra
uterine walls
uterus
vasopressin
veins
venules
villi
vitamin K
voluntary muscles
yolk sac
zygote

Chapter 8
BEHAVIOR AND ECOLOGY

OVERVIEW

The two areas included in this chapter take an even broader view of living things. Starting with animal behavior, we explore the way that animals interact with their environment as an individual behavioral matter, specifically how they learn and react to their surroundings. Following this, we outline what we know about how animals communicate, move, feed, and interact socially and, finally, the subject of circadian—literally, about a day—rhythms. The second section is involved with an even broader view than the individual behaviors covered in the first section. Taking a world-view, we outline what the student may be tested on, beginning with the relationships between members of a population and then their involvement with other species at a community level. We outline the different kinds of communities within the biosphere and, by including the abiotic factors, we study the interactions between the biosphere, the hydrosphere, the lithosphere, and the atmosphere on this planet in different regions and climes called biomes. From there, a world-view, literally, is outlined in summing up the relationships of all living things to each other and the planet and the changes they—living things and the planet—undergo.

ANIMAL BEHAVIOR

This is an area not specifically covered in any of the other areas in this book but actually brings together several factors in describing the way life expresses itself, given the interaction of all those other factors. In addition to learning, the areas covered alternately relate to the individual, groups, and their interactions. Some of the areas covered involve the study of influences outside the individual; others involve those that "come with the package." The argument persists as to whether nature or nurture dominates life. Do we come with a prepackaged set of formative influences, or do the environment and our response to it shape everything?

LEARNING

This is usually defined as acquiring information or a behavior that was not previously a part of the organism and tends to focus on outside influences. We now believe that the genes play an important role in the behavior of an individual and the individual's relations with the environment, including other individuals, for example, as in the case of reproductive success or survival fitness. Evolution and natural selection are seen to enhance these genetically conferred traits.

Conditioning

Conditioning is defined as exhibiting a behavior as a result of a response to a particular set of conditions, known or unknown, usually called a stimulus. Care should be taken not to "run around in circles" on the matter of stimulus/response. A stimulus is anything that results in activity; a response is the result produced.

A. Classical

Classical conditioning, or associative behavior, is of course, given in Pavlov's classical "salivating dog" experiment and is taken as responding to a substitute stimulus, rather than the original or expected stimulus.

B. Operant

Operant conditioning is more a trial-and-error kind of conditioning. The closer an individual gets to the desired response with positive reinforcement, the more the individual will repeat the response. This occurs in animal training but can also be seen in an animal that spends initial energy on revisiting a feeding site that has yielded success in the recent past.

Habituation

Habituation involves a behavior in which unproductive results are soon disregarded and the stimulus that causes them is also, eventually, disregarded.

Insight

Without any previous experience with a situation, an animal performs a successful behavior, most likely influenced by previous behaviors.

Observational

Individuals copy the behavior of another without having experienced the stimulus previously that produces that behavior.

INNATE

Behaviors, or the potential for them, that already exist inside the individual are called innate.

Imprinting

This behavior is an example of one that appears to involve a "critical period." It occurs only if the right stimulus occurs during a particular time. Examples would be salmon returning to the precise site of their birth or as in Lorenz's studies with goslings.

Fixed action pattern

These follow a regular, unwavering pattern that is initiated by a specific stimulus and carried out even in the absence of the ability to complete the original intent.

Instinct

Instinct is the classic innate behavior that has no other explanation other than that it is inherited. One such behavior appears to deal with time and is known as circadian rhythm. Internal clocks or cycles appear to have no source other than an instinctual interaction between the environment and chemical promoters in the individual or by genetic factors.

MOVEMENT

Animals have particular ways in which they move. Three specific ways have been identified and they involve very specific actions according to the organisms and the environment in which they reside.

Kinesis

Kinesis is a random change in speed and/or direction of an organism in response to a stimulus. They will, for example, speed up in an unfavorable environment and slow down in a favorable one.

Taxis

Taxis is a very specific, directed response to a stimulus, in the direction of the stimulus if it is favorable and away from the stimulus if it is unfavorable.

Group

Group movement, better known as migration, is the wholesale movement of animals that is usually brought on by seasonal influences, in particular the availability of food.

COMMUNICATION

Communication is usually observed within a species either to recognize one's own kind for mating or for social behavior of one kind or another, one of which may be the avoidance of a nearby predator.

Chemical

Chemical communication uses pheromones, which trigger a response when smelled or, in some cases, when eaten. In the case of bees, the workers eat a pheromone, and it prevents them from developing reproductive ability.

A. Releaser pheromones
Releaser pheromones elicit a behavior when experienced.

B. Primer pheromones
Primer pheromones elicit a physiological change when experienced.

Visual

Visual communication is usually observed during courtship or aggressive behavior when defending territory or establishing pecking order. In some, visual displays by males are releasers for reproductive behavior in females.

Auditory

Auditory communication may be for definition of territory; for species recognition, as in the case of a mating behavior; or for alerting behavior, as when a predator is nearby.

Tactile

Tactile communication is a common behavior in bonding, infant care, and mating. In bees, it is even used to locate successful food sources.

FEEDING

Feeding behaviors that organize the search for food have been observed and appear to impact the survival of the species and have, therefore, great impact and importance for the individual, even as the individual may relate to a group.

Groups

Groups of individuals have evolved several successful means of feeding or finding food.

A. Herds
Herds use this kind of group behavior when feeding as a means of defending against predators by shielding each other, when attempting to avoid a predator by becoming hidden from view, or simply by

being alert to the presence of a predator. A member on one end of the herd may not be aware of a predator approaching the other end of the herd.

B. Packs
Packs engage in group hunting, which is most often observed when smaller species attack a larger one.

Search image
Animals will seek food based on the success previously experienced or on images they possess, however limited, of the target previously experienced. A measure of habituation obviously contributes to the success of this feeding behavior.

SOCIAL

Individuals behave in two different settings: in a group or on their own. Eventually, they will make contact with other organisms within their species or outside their species. The most successful interactions survive, as do the individuals or the groups who exhibit them.

Agonistic
Agonistic behavior is aggressive/submissive behavior that actually has great survival value, since it is most often quite ritualized and, therefore, actually reduces negative aspects of aggression.

Dominance
Established hierarchies, or "pecking order," also minimize conflict within the group.

Territoriality
Territoriality is a behavior that ensures a food supply, a place for reproductive activities, and successful rearing of infants.

Altruistic
Altruism is seemingly unselfish behavior that actually increases the fitness of the group by ensuring survival of the species and parental genes when defending another.

ECOLOGY

The word ecology comes from the root *ecce,* meaning home or place. If, in several instances, we have talked about the big picture, this is the biggest of all. All of the material discussed so far culminates in the study of our environment or ecology, the study of "this place."

BIOSPHERE

The biosphere is the realm of living things, along with the hydrosphere, lithosphere, and atmosphere, that make up our world.

The Earth

The earth includes the four "spheres" and the ways in which they interact.

Major Biomes

Under the influence of latitude or elevation and various abiotic factors, the earth is divided into large areas known as biomes or areas of distinctly different living things. Beginning with the most inhospitable because it is the coldest, we begin at the "top" of the world. While reference will be made only to latitude as we progress toward the equator, one can also find similar conditions when descending a mountain.

A. Tundra

The northernmost biome is tundra, which contains rapidly maturing plants, a permanent layer of ice under the soilm and mostly small homeotherms.

B. Taiga

Taiga is less forbidding. It nonetheless has long cold winters; larger plants, such as stunted conifers, exist; and there are more large animals.

C. Temperate deciduous forests

Temperate deciduous forests contain larger trees that lose their leaves in dry—usually winter—seasons, proliferate heterotrophs and autotrophs, and have warm summers and colder winters.

D. Grasslands

Grasslands feature hot summers, cold winters, unpredictable rainfall, mostly grassy autotrophs, and a fairly wide variety of wildlife.

E. Deserts

Deserts contain drought-resistant plants, low rainfall, a very dry climate, small homeotherms, and proliferate reptilia.

F. Tropical rain forests

The tropical rain forests are the most proliferate of all biomes for living things, containing high rainfall and temperatures and poor soil.

ECOSYSTEM

An ecosystem is smaller than a biome and is most often a self-contained area, such as a lake or a tidal pool.

Abiotic factors

Abiotic factors are nonliving factors that include air, water, topography, geology, and sun. The interactions of these include things like temperature, humidity, wind, and erosion.

Biotic factors

Biotic factors are living things in an ecosystem and the way they interact.

A. Producers

Autotrophs are known as producers since, in the presence of the ultimate source of energy—the sun—they provide most of the biomass.

B. Consumers

As the name suggests, these consume biomass but are not capable of producing or using sunlight directly to create a food source, as the autotrophs are. Consumers are usually divided into at least primary consumers, whose primary source of nutrition is producers, and secondary consumers, whose primary source of nutrition is another consumer.

1. Herbivore—designed to metabolize producers exclusively.
2. Carnivore—designed to metabolize other consumers exclusively.
3. Omnivore—designed to metabolize both producers and consumers.

C. Decomposers

At some point, the complex molecules that living things construct from their nutrients must be broken down lest the environment run out of raw materials that some organisms need to carry on life. This is the role of the decomposers.

FUNCTIONING

The biotic world functions at several levels which are outlined here.

Food chain

The food chain consists of a linear interaction between organisms in which the starting point is usually a plant and is consumed by the next organism that is consumed by the next organism and so on until the chain ends. It can be portrayed vertically in what is called a food pyramid that more accurately portrays the amount of biomass at each level in the chain (called a trophic level).

Food web

The food web is several food chains interacting. For example, a mouse might not just eat corn, but it might eat nuts or grains as well.

Community

A community is a group of populations living in the same area.

A. Habitat

A habitat is the place where organisms usually live.

B. Niche

A niche consists of all the living and nonliving sources that contribute to an animals survival—or, more accurately, a species survival—as a result of successful achieving of equilibrium with these factors by the animal (species). A niche is an organism's "job," or what it does in an ecosystem.

SYMBIOSIS

Relationships where animals are in close proximity to each other and interact as a result. Literally, it means "living together."

Camouflage

Camouflage is the successful concealment from view of an organism, usually from its predator.

Commensalism

Commensalism is an interaction between organisms where one benefits and the other is not harmed.

Mimicry

Mimicry is apparently similar to camouflage in that a protective appearance has evolved, but this might be displayed rather than concealed, since it might resemble a predator of an attacker or, at the very least, a more aggressive or distasteful organism than the attacker planned on encountering.

Mutualism

Unlike commensalism, both organisms benefit in mutualistic interaction.

Parasitism

Parasitism must strike a delicate balance between the parasite and the organism off of which it lives, mainly the host. The parasite-host relationship is one in which one organism is helped; the other—the host—is harmed but not killed.

Predator/Prey

The killing of the prey benefits predator.

POPULATION STUDIES

This very important area of science is applied primarily to the study of human populations, although the principles relate across all lines of living things. Generally, the areas of study seek to determine the nature of the population, particularly its health or equilibrium with the environment.

Size

The number (N) of individuals in a population.

Density

N per unit of volume—not area since some organisms occupy a three-dimensional cross-section of the niche.

Dispersion

How the population is distributed.

Age structure

The N of each age group, diagrammed along a vertical line that separates male and female members of the population on either side, results in specific shapes that can be interpreted as population trends.

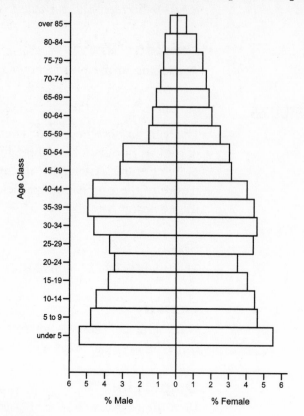

Mortality

Mortality is the survival of individuals throughout their lifetime.

Growth

Two organisms and their life strategies emerge from our studies of populations. Their growth is the result of these strategies as the impact the environment, and vice versa. The r-selected species have a lifestyle of rapid invasion, rapid increase—from birth rates that involve many small offspring that mature quickly—and rapid decline. On the other hand, K-selected species numbers remain at the carrying capacity and have few, relatively large offspring who require much care.

A. Biotic potential

Maximum growth rate of a population under ideal conditions for all the factors just discussed.

B. Carrying capacity

Indicated by the use of K, it is the maximum N that can be sustained by a particular habitat.

C. Limiting factors

Those factors in a habitat that prevent a population from reaching its maximum N or biotic potential.

1. *Density dependent*—limiting factors that depend on the size of the population and have greater and greater effect on the population as the population grows such as competition for food, dispersal of wastes and reproductive behaviors.

2. *Density independent*—those factors that do not have a differential effect on the population, depending on the size of the population. Natural disasters and weather/climate extremes are among the density-independent limiting factors.

D. Exponential

Exponential growth occurs when the birth rate is >0, making the increase in population exponential. When graphed it typically follows a J-shape as the slope continues to increase towards a 90° incline.

E. Logistic

Logistic growth occurs when limiting factors restrict the growth rate to the available resources (carrying capacity). It is represented by an S-shaped or sigmoid curve that typically plateaus when the population reaches equilibrium with the carrying capacity.

SUCCESSION

As the name suggests, it is the change in the make-up of populations in a community over time ending, with a climax community for that biome.

Primary

Primary succession is the first instance where an area will experience succession; the area has never before experienced living things, as in immediately after a lava flow—usually extensive in size—cools.

Secondary

Secondary succession occurs when an area previously occupied by living things is completely destroyed, as in a forest fire or a flood that complete washes out an area.

MULTIPLE CHOICE QUESTIONS

Questions 1-4 refer to the following relationships:
(A) predator-prey
(B) commensalism
(C) mutualism
(D) parasitism
(E) camouflage

1. A cow produces milk daily after grazing in a large pasture owned and maintained by the farmer.

2. Remora fish attach themselves to the underside of a shark, and while the shark feeds, the remora darts out for shreds of the shark's food that are left floating in the water.

3. A walking stick resembles the branches it resides on.

4. Lacking a digestive system, tapeworms live in the intestines of their host.

5. Which of the following animals in the food chain has the largest biomass?

    (A) plants

    (B) grasshopper

    (C) mouse

    (D) owl

    (E) snake

6. Which of the following is an instinct?

    (A) caring for offspring

    (B) habituation

    (C) operant conditioning

    (D) associative behavior

    (E) observational learning

7. In the graph below, the curve indicates the

   (A) biotic potential

   (B) carrying capacity

   (C) density-dependent effect

   (D) density-independent effect

   (E) sigmoid curve

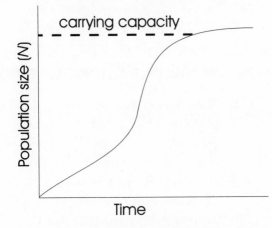

8. All of the following are associated with mating behavior except

   (A) aggressive behavior

   (B) releaser pheromones

   (C) search image

   (D) territoriality

   (E) visual communication

9. Which of the following biomes is characterized by the driest conditions?

   (A) tundra

   (B) taiga

   (C) temperate deciduous forest

   (D) grasslands

   (E) tropical rain forest

10. Which of the following is NOT a means of animal communication?

(A) auditory

(B) chemical

(C) instinctual

(D) tactile

(E) visual

EXPLANATIONS OF ANSWERS FOR MULTIPLE CHOICE QUESTIONS

1. **The correct answer is (C).** The relationship indicated in this question benefits both individuals and so would be a case of mutualism. The next closest would be commensalism, but, in this case, both, not one, are benefited.

2. **The correct answer is (B).** This relationship is the case where commensalism is illustrated. While the shark is not harmed by the relationship, neither is it benefited. Only the remora benefits from the free transportation and the ready supply of food.

3. **The correct answer is (E).** Since the walking stick resembles the branch on which it lives, it is hidden from the view of predators as it camouflages itself by such an adaptation.

4. **The correct answer is (D).** Using another species for its survival by taking some part of its host's means of survival, the tape worm is acting like a parasite. Due in part to its own lack of a digestive system, the tapeworm must achieve a balance, through its reproductive behavior, with the host lest the tapeworm kill the host, which would mark the demise of the tapeworm itself. In a parasite-host relationship, the parasite must reach a balance with the host that does not kill the host.

5. **The correct answer is (A).** If we construct the food chain with the member we are given as a food pyramid, we see that it must begin with the plants in a position at the bottom and widest part of the pyramid, which indicates that it represents the largest biomass. As we continue up the pyramid, or chain, the amount of material—which houses the main ingredient in the picture, that of energy—decreases dramatically between levels. The end result is that 10 percent of the energy, on average, of any one level is passed on to the next level. The owl, which we might find at the top of this food pyramid, has the least amount of biomass. Next comes in the snakes it consumes with more biomass, just below the owl on the pyramid. More so the mouse,

even more so the grasshopper until we reach the bottom where the largest amount of biomass is found in the plants.

6. **The correct answer is (A).** Choices (B) through (E) are all forms of learned behavior and involve a necessary interaction with the environment to take place. Caring for an infant seems to be one of those behaviors that is instinctual in most species. Habituation is the active disregarding of unnecessary stimuli. Operant conditioning is the trial and error kind of learning associated with acts such as training. Associative behavior is the classical form of conditioning where an animal associates a set of behaviors with a successful result, usually after repeated encounters. Observational learning is associated with mimicking another's behavior after close observation.

7. **The correct answer is (B).** The entire graph is known as a sigmoid graph and represents changes in a population over time. The plateau indicates equilibrium between the organisms and the environment. A drop in the graph would indicate increasing mortality; a rise would indicate increasing birth rate. While biotic potential is a seemingly attractive choice, this graph is an indication of the reality of the population. Another line would be needed to indicate density relations, such as the relation between r-selected species and K-selected species.

8. **The correct answer is (C).** Search image has to do with a feeding or foraging behavior in an organism as it seeks food sources of which it has retained a somewhat abbreviated image. All of the other behaviors impact mating behavior by establishing dominance, triggering a behavior by chemical means, establishing a territory of ownership, or by a display that the opposite sex recognizes or which triggers reproductive behavior in the opposite sex.

9. **The correct answer is (A).** From the list of biomes given, the tundra classifies as the driest. Although it is covered with water, this water is frozen and therefore unavailable in the way we would hope water would be available for any plant or animal use. The taiga, with a period during which the water is frozen, also has a time when it is liquid and available for use. Temperate deciduous forests are more wet, and tropical rain forests are the wettest of all. Grasslands fall somewhere in between the taiga and temperate biomes. If there were more moisture in the grassland biome, it could support larger plants. On the other hand, it must have enough moisture to support the grassy plants throughout a growing season.

10. **The correct answer is (C).** Animals communicate through a variety of means, the four major ones being listed here. Instinc-

tual is not a means of communicating; rather, it is how the organism responds to environmental needs without appearing to learn these responses. Sound, chemical, touch, and sight are all ways that animals communicate with each other.

VOCABULARY

associative learning
autotroph
behavior
biomes
biosphere
carrying capacity
circadian rhythms
classical conditioning
climax community
commensalism
community
consumer
critical period
decomposer
density-dependent factors
density-independent factors
ecology
ecosystem
exponential growth
fauna
flora
food chain
food pyramid
food web
herbivore

heterotroph
imprinting
insight
instinct
K-strategists
learning
logistic growth
mutualism
niche
operant conditioning
parasitism
pioneer community
population
population density
primary consumer
producer
r-strategists
reasoning
secondary consumer
sere
succession
trophic

Chapter 9
LABORATORY

OVERVIEW

In the absence of the possibility of having students actually perform labs as a part of the AP Biology Exam, it tests your knowledge of the expected results from the labs they present. The literature itself outlines the objectives for each lab and what you should be able to derive from the information given on the exam. One problem is that not all schools perform every experiment in their AP curriculum. This is understandable given the wealth of information that must be covered in addition to performing the labs. Therefore, the student may find our review book helpful in providing as much information in this chapter and others as the student should reasonably be expected to know in order to be successful on this part of the test. We will outline what the lab is about, what you need to know to successfully deal with the lab, and any tips that might be helpful. First, we will outline the labs, and the last section will provide tips on successfully dealing with this part of the exam. In the "First Things" section, we briefly describe what the student will do in the lab, despite the fact that the student may not actually have the opportunity to do the lab. In "The Basics," we indicate what the student should know before doing the lab and what the expected outcomes are, and finally, in the "Last Things" section, we provide any additional information the student know.

LABORATORY 1—DIFFUSION AND OSMOSIS

First Things

The student will demonstrate a competency in investigating the processes of diffusion and osmosis and the effect of solute concentration on water potential in plants.

The Basics

In order to be successful on this lab, the student should understand (a) the mechanisms of both processes, (b) water potential, (c) differentially (selectively) permeable membranes and their effect on the two processes, (d) molarity, and (e) the relationship between water potential and solute concentration. Following this lab, the student will be able to (a) measure water potential of a solution, (b) determine the osmotic concentration in living plant tissue, (c) relate the effects of water gain or loss in living tissue, and (d) relate osmotic pressure to solute concentration and water potential.

LAST THINGS

The student should be familiar with the fact that the free energy in water is known as its water potential. The terms *hypotonic* and *hypertonic* should be well known as well as *turgor pressure*. Finally, the student should be familiar with the basics of the movement of water across a membrane and its effect in animal and plant tissue.

LABORATORY 2—ENZYME CATALYSIS

FIRST THINGS

The student will measure the action of the enzyme catalase on hydrogen peroxide and the rate at which it is converted to water and oxygen gas.

THE BASICS

In order to be successful on this lab, the student should understand (a) enzymes, their structure, and how they act; (b) free energy; and (c) initial reaction rates of enzymes and the conditions that affect them. After performing this lab, the student should understand how to (a) measure the effects of temperature, pH, and enzyme/substrate concentration on enzyme-mediated catalysis; and (b) explain the effect that environmental factors have on enzyme-catalyzed reactions.

LAST THINGS

The student should be familiar with enzymes and how they act and how, if they become denatured, they change their shape and thus their ability to work.

LABORATORY 3—MITOSIS AND MEIOSIS

FIRST THINGS

The student will perform two labs, one on mitosis and one on meiosis.

A. Mitosis

The student will calculate how long a cell spends in the various phases of mitosis and compare animal mitosis with plant mitosis.

B. Meiosis

The student will investigate the stages of meiosis using chromosome models. This includes crossing over and recombination. The student will investigate the life cycle of the fungus *Sordaria fimicola,* the arrangement of ascospores, crossing over, and the gene control of spore color.

THE BASICS

In order to be successful on this lab, the student should understand (a) the events in mitosis, (b) the differences between mitosis and meiosis, and (c) the events in gametogenesis. After performing this lab, the student should understand (a) the relative duration of cell cycles; (b) the mechanisms of crossing over, segregation, and independent assortment; and (c) how to calculate the distance of a gene from the centromere.

LAST THINGS

The student should be familiar with mitosis and meiosis and how it leads to genetic variability.

LABORATORY 4—PLANT PIGMENTS AND PHOTOSYNTHESIS

FIRST THINGS

The student will separate pigments using chromatography and measure the rate of photosynthesis through the reduction of the dye DPIP.

THE BASICS

In order to be successful on this lab, the student should understand (a) chromatography, (b) photosynthesis, (c) function of plant pigments, and (d) the effect of light on photosynthesis. After performing this lab, the student should understand (a) how to separate plant pigments, (b) how to measure photosynthetic rates, (c) the effect of temperature and light on the rate of photosynthesis, and (d) how to explain the variability of photosynthetic rates.

LAST THINGS

The student should be familiar with paper chromatography, how to calculate the rate of pigment migration, how light and temperature affect photosynthesis, and how to design an experiments to test these three variables.

LABORATORY 5—CELL RESPIRATION

FIRST THINGS

The student will measure oxygen consumption during respiration in germinating or nongerminating peas.

THE BASICS

In order to be successful on this lab, the student should understand (a) how a respirometer works and (b) the process of metabolism in a living organism. After performing this lab, the student should understand how to (a) test the effects of temperature on peas in a controlled experiment, (b) calculate the rate of respiration, and (c) relate gas consumption and production to the respiration rate.

LAST THINGS

The student should be familiar with the formula for cellular respiration and how a respirometer works.

LABORATORY 6—MOLECULAR BIOLOGY

FIRST THINGS

The student will investigate some basic principles of genetic engineering.

THE BASICS

In order to be successful on this lab, the student should understand (a) gel electrophoresis, (b) principles of bacterial transformation, (c) proper cell preparation, (d) the role of plasmids in genetic engineering, (e) how restriction enzymes function, and (f) transfer of antibiotic resistance. After performing this lab, the student should understand how to (a) use plasmids in genetic research, (b) use restriction enzymes, (c) calculate transformation efficiency, (d) use multiple experimental controls, and (e) use DNA fragments of known size to calculate the size of unknown fragments.

LAST THINGS

The student should be familiar with how bacteria incorporate foreign DNA, how restriction enzymes function and electrophoresis and how bacteria are "transformed."

LABORATORY 7—GENETICS OF DROSOPHILA

FIRST THINGS

The student will use fruit flies to investigate genetic crosses.

THE BASICS

In order to be successful on this lab, the student should understand (a) chi-square analysis and (b) the life cycle of diploid organisms. After performing this lab, the student should understand how to (a) investigate independent assortment of two genes, (b) use a multi-generational study to investigate the interactions between the two genes, and (c) analyze the data from chi-square analyses.

LAST THINGS

The student should be familiar with Punnett squares and multigenerational study techniques.

LABORATORY 8—POPULATION GENETICS AND EVOLUTION

FIRST THINGS

The student will learn the Hardy–Weinberg law and the relationship between evolution and gene frequency.

THE BASICS

In order to be successful on this lab, the student should understand (a) how natural selection alters gene frequency, (b) the Hardy–Weinberg equation, and (c) the effects of allele frequency on natural selection. After performing this lab, the student should understand how to (a) calculate the frequency of alleles in the gene pool, (b) use the Hardy–Weinberg law to do this, and (c) discuss deviations from Hardy–Weinberg.

LAST THINGS

The student should be familiar with the Hardy-Weinberg law and know how to calculate allele frequencies and evolution.

LABORATORY 9—TRANSPIRATION

FIRST THINGS

The student will measure transpiration under varying laboratory conditions and how the structures in the plant stem and leaf relate to this.

THE BASICS

In order to be successful on this lab, the student should understand (a) how water moves in plants, (b) the role of transpiration in moving water in a plant, and (c) plant physiology. After performing this lab, the student should understand how to (a) test the role that environmental variables play in the rate of transpiration, (b) make a thin section of tissue, and (c) identify the cells in plant vascular tissue.

LAST THINGS

The student should be familiar with transpiration, plant physiology, and how to make thin sections of tissues.

LABORATORY 10—PHYSIOLOGY AND THE CIRCULATORY SYSTEM

FIRST THINGS

The student will, under a variety of conditions, learn how to measure and analyze blood pressure, measure pulse rate, and determine the effect of temperature on the heart beat of a water flea.

THE BASICS

In order to be successful on this lab, the student should understand the relationship between temperature and the rates of physiological activities. After performing this lab, the student should understand how to (a) measure heart rate and blood pressure, (b) describe the effect of body position on heart rate and blood pressure, (c) analyze cardiovascular data, (d) explain how exercise changes heart rate, and (e) discuss the effect temperature has on heart rate.

LAST THINGS

The student should be familiar with how to measure pulse and blood pressure and should know that heat increases physiological activity.

LABORATORY 11—HABITAT SELECTION

FIRST THINGS

The student will observe the behavior of an insect and design an experiment to investigate its responses to changing environments. The student will also observe mating behavior.

THE BASICS

In order to be successful on this lab, the student should understand (a) organism distribution with respect to resources, (b) kinesis and taxis, (c) how environmental factors affect habitat selection among organisms, and (d) how to describe different mating behaviors. After performing this lab, the student should be able to (a) design an experiment to investigate an organisms responses to environmental variables, (b) measure the effects of various environmental variables on habitat selection in a controlled experiment, and (c) describe the different types of insect mating behavior.

LAST THINGS

The student should be familiar with the variables that affect an organisms habitat selection.

LABORATORY 12—DISSOLVED OXYGEN AND AQUATIC PRIMARY PRODUCTIVITY

FIRST THINGS

The student will analyze the dissolved oxygen content in water and the productivity of laboratory cultures relative to the intensity of light.

THE BASICS

In order to be successful on this lab, the student should understand (a) the carbon and oxygen cycles in nature, (b) primary productivity in an ecosystem, (c) solubility of gases in water, and (d) how photosynthesis, respiration, and dissolved oxygen affect primary productivity. After performing this lab, the student should understand (a) how to measure primary productivity and (b) the effect that light and inorganic nutrients have on primary productivity.

LAST THINGS

The student should be familiar with the carbon and oxygen cycle; the effect that temperature, salinity, and photorespiration have on the solubility of dissolved gases; primary production; and the difference between gross and net productivity.

LAST THINGS

Here, finally, are a few things to be aware of. In this area, students will be tested either on experimental design or analysis. If you must construct a graph, don't forget labels—use the x-axis for independent variables and the y-axis for dependent one. Where appropriate, connect the dots and provide values along the axes clearly and regularly. When you design an experiment, we want to distinguish between the dependent and the independent variables. Identify the experimental values and the control values, and remember that every experiment tests only one independent variable at a time. Organisms must be the result of a random sample. Describe the procedure, the expected results, and why you expected those results based on applicable biological principles.

Chapter 10
WRITING THE ESSAY

OVERVIEW

In this chapter, we will offer advice on writing as good an essay as you can. It may be curious to discover a mostly language arts skill on a science test. The fact is, in science, as in other endeavors, we have to be able to communicate our findings, our successes, and our failures in every endeavor in life—not the least of which is science. When one endeavors to write coherently and cohesively, one needs to organize thoughts and information in much the same way that an experiment must be organized. Someone wrote recently, "How do I know what I think unless I see it in writing?"

Perfect writing does not come with the first effort; it must be revised and revised and revised. Just as thoughts do not easily flow onto the paper, a scientific experiment, with the variables poorly understood, will also not make sense. Here, then, are a few thoughts and guidelines on writing a good essay. At the end, we reproduce several past essays and the scheme used to correct them. We feel that the more of the items ETS gives points for and the student includes the higher the score the readers will award. Let the rubric serve as a checklist for each essay. Encouragement is given to the student to practice writing essays from the checklists we give.

GUIDELINES

Many references caution you to read the question carefully. We agree. Many also urge you to brainstorm: write down anything that comes to mind on the subject of what the question appears to be asking—more on that later. Again, we agree. Next, the student is often urged to organize the list of thoughts that resulted from brainstorming in a more or less outline form. If a related thought or piece of information pops into your head at anytime during the writing process, jot it down. Also, we suggest you jot down the results of your brainstorm in a place where the readers will see it. Your attempt to let the information flow will be noticed. It may not be worth formal points, but the clear attempt to generate information and the subsequent organizing will cause readers to take more careful notice of your finished product.

We don't suggest a strict outline with Roman numerals, letters, and numbers but in the following way: take the first item on the list and number it "1." Search the list for any other items that relate to this item and label them with a "1" also. Once all the ones have been identified, go back and label the next unlabelled item with a "2." Likewise, search the list for related items, label them with a "2" and so on. Once all the items are organized in this way, go back and consider all the ones as the information that will be presented in the first paragraph. Decide which "1" should come first and label it as "1a," the second as "1b," and so on. Repeat this with all the other numbered items. The good news is that the process organizes the information so that the essay will almost write itself and that describing the process takes far longer than doing it. We want to make one suggestion not found in most references we have read and that is, before beginning the writing process go back and reread the question. Often, we start writing an essay with what, in our mind, we think the question is asking. It is better to make one more effort to clarify what the question is really asking.

In writing the essay, a few simple suggestions are in order. For example, organize your essay the way you would an experiment. In fact, you are writing about science experiments so it is of value to develop your ideas one step further; don't be verbose, but don't be afraid to be a little more wordy than you think might be needed.

You have about 22 minutes to plan and write your essay, but plan for 20 minutes and use the extra two minutes as a cushion. Be as clear and as logical as you ever have been. Since there are four essay questions on the exam, each essay is worth 25 percent of your overall score. Use the practice essays to hone your skills and your sense of time. Try the practice essays, and develop an essay mentality under essay-like conditions. Time yourself, and stick to the time. Like many references urge, and we concur: PRACTICE, PRACTICE, PRACTICE.

Here, then, are several essay questions. We urge the student to read the questions, become familiar with the correction scheme, and practice.

The "STANDARDS" designation indicates the point value allowed for each answer that is preceded by an underline. The indication of "5 points" or "6 points" shows the total number of points possible in each section despite there only being, for example, "MAX 3," which means each section is awarded a maximum of 3 points. The AP folks are allowing 5 chances [or 6 chances] to get three points.

QUESTION #1

1. Explain what occurs during the Kreb's (citric acid) cycle and electron transport by describing the following:

(A) The location of the Kreb's cycle and electron transport chain in the mitochondria

(B) The cyclic nature of the reactions in the Kreb's cycle

(C) The production of ATP and reduced coenzymes during the cycle

(D) The chemiosmotic production of ATP during electron transport

STANDARDS:

3 points maximum for each of the four sections.

1 point for any of the following answers:

LOCATION

___Description of internal structure (compartmentalization) of mitochondrion

___Kreb's in matrix (inner or M compartment)

___Kreb's enzymes mostly dissolved in matrix

___ETS in cristae (inner membrane)

___ETS components are embedded in the inner membrane

5 points—Answer a maximum of three.

CYCLIC NATURE OF KREBS

___Acetyl CoA (C2) starts Kreb's

___C2 joins with OAA (C4) to form citric acid (C6)

___2 CO2 removed during Kreb's

___OAA is recycled or overall cycle concept (diagram OK)

___Position of Kreb's in aerobic respiration (2 cycles/glucose, uses products of glycolysis)

5 points—Answer a maximum of three.

PRODUCTION OF ATP AND REDUCED COENZYMES

___1 ATP/cycle

___NADH and/or FADH2 formed (or NADH2, NADH + H+, NAD red)

___Amount of NADH (3) and/or FADH2 (1) per cycle

___ATP formed from released energy; substrate level phosphorylation

___ATP specific reaction: Succinyl CoA -> Succinic Acid; GTP —> GDP

ADT -> ATP

___NADH or FADH2 formed by H or e-

6 points—Answer a maximum of three.

CHEMIOSMOTIC PRODUCTION OF ATP
___Electron transfer (redox) through carriers; O_2 final acceptor
___Gradient drives ATP formation or battery (electrochemical, charge separation, etc.)
___Protons pumped to inter-membrane space;
    proton carriers alternate w/electron carriers;
    charge separation; gradient established;
    lowers pH in inter-membrane space;
    + between membranes;—matrix
___Inner membrane impermeable to H+ except for and/or proton channel (ATP synthetase; ATPase) is permeable
___ATP synthetase structure F0 + F1
___Specifics of number of ATPs formed (ETS = 32 ATP; 2 ATP/FADH2; 3ATP/NADH)

QUESTION #2

1. Discuss the adaptations that have enabled flowering plants to overcome the following problems associated with life on land.

    (A) The absence of an aquatic environment for reproduction

    (B) The absence of an aquatic environment to support the plant body

    (C) Dehydration of the plant

STANDARDS:

4 points maximum for each of the following sections.

1 point for any of the following:

A. ABSENCE OF AN AQUATIC ENVIRONMENT FOR REPRODUCTION:
___Flowers—attraction for insects—shape, color, smell, chemical, nectar.
    Mimicry for pollination (co-evolution)
___Timing of reproduction—Male
___Microspores—pollen/reduced gametophyte
___Lack of motility of gamete—pollen grain modification for transport, e.g., light weight/structure
___Pollination—transport of male gametes, wind, insects (self-pollination); female
___Reduced gametophyte (in megaspore or megasporangium)
___Protected gametophyte—embryo inside ovary, carpel, pistil

___Evolution of seed
___Fertilization—internal—pollen tube, endosperm
___Fruit and seed dispersal
___Seed dormancy

11 points—Answer a maximum of four.

B. ABSENCE OF AN AQUATIC ENVIRONMENT TO SUPPORT THE PLANT BODY:

___Stem support
___Root—anchorage
___Vascular tissue—xylem fibers, tracheids, vessels, heartwood, dead tissues, phloem fibers or vascularization
___Vines, tendrils
___Cell wall—lignin, cell wall support, cellulose
___Cambium—secondary thickening
___Sclerenchyma—whole wall support
___Collenchyma—corner wall support
___Prop, buttressed roots
___Turgor pressure

10 points—Answer a maximum of four.

C. DEHYDRATION OF THE PLANT:

___Root hair, absorption
___Cuticle, wax, acellular
___Bark—suberin, cork
    Scales—bud protection
    Sepal/petals—floral part protection
___Seed coat/pollen grain wall
___Stomates—function to control water movement
___Xylem—water transport
___Leaf/stem/root modifications (2 points maximum)
    surface area reduction in desert plants/succulents
    stomates under surface
    leaf rolling
    hairs and trichomes
    interlocked epidermal cells
    hypodermis
    cortex—water storage or retention
    loss of leaves/abscission layer
___CAM/C4 plants—modified stomate functions
___seed dormancy (if not mentioned in part A)

QUESTION #3

1. Describe the differences between the terms in each of the following pairs.

   (A) Coelomate versus acoelomate body plan

   (B) Protostome versus deuterostome development

   (C) Radial versus bilateral symmetry

2. Explain how each of these pairs of features was important in constructing the phylogenetic tree shown below. Use specific examples from the tree in your discussion.

STANDARDS:

Four points maximum for each section.

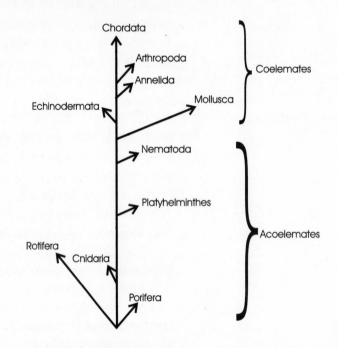

Phylogenetic Tree

2 points for any of the following answers:

A. (1) COELOMATE VS. ACOELOMATE

1-Coelomate: internal body cavity lined with mesoderm (not sufficient to say: "true body cavity")

1-Acoelomate: lacking internal cavities altogether or having:

    a pseudocoelom (Nematoda and Rotifera)

    a spongocoel (Porifera)

    mesoglea (Cnidaria)

    a solid layer of mesoderm (Platyhelminthes)

4 points—Answer a maximum of two.

(2) PROTOSTOME VS. DEUTEROSTOME DEVELOPMENT

1–Protostome: mouth develops near/at the blastopore or anus forms secondarily (later),

OR featuring: spiral cleavage (micromeres between macromeres);

determinate/mosaic development (blastomere fate is established at very early stages of development);

mesoderm from cells that migrate into the blastocoel near blastopore schizocoelous coelomation (internal split in solid wedge of mesoderm that is independent of gut);

trochophore larva;

1–Deuterostome: anus develops near/at the blastopore or the mouth forms secondarily (later),

OR featuring: radial cleavage (micromeres directly above macromeres);

indeterminate/regulative development (blastopore fate is variable and not established until late in development);

mesoderm arises from outpocketings of the gut;

enterocoelous coelomation (outpocketings of gut);

dipleurula larva

4 points—Answer a maximum of two.

(3) RADIAL VS. BILATERAL SYMMETRY

1–Radial: several planes passing through the long or central axis can divide the organism into similar parts.

1–Bilateral: (only) one plane passing through the long axis divides the organism into similar right and left sides— exhibits cephalization.

1–Echinoderms: bilaterally symmetrical larvae but appear to have radially symmetrical adult forms.

6 points—Answer a maximum of two.

B. PHYLOGENETIC TREE

1–for examples of contrasting pairs (phyla or organisms) using terms from above; answer here or in part A.

1–for using above terms in explanation of why phyla are in separate groups (or separate branches) of the tree.

1/1–Body symmetry (cephalization) permits separation of Porifera and Cnidaria (radially symmetrical) from other phyla (bilaterally symmetrical).

1/1–Coelomation permits separation of Platyhelminthes, Nematoda, and Rotifera from other phyla above Cnidaria: flatworms are acoelomate, whereas those other than nematodes and rotifers are coelomate.

1/1-Origin of the mouth and anus permit separation of
Echinodermata and Chordata (deuterostomes) from
Arthropoda, Annelida, and Mollusca (protostomes).
[Some include Platyhelminthes, Nematodes, and Rotifers as
protostomes.]
1-Nematodes and rotifers are grouped separately because both
are pseudocoelomate.
1-Phylogenetic trees based taxonomic relationships on
homologous structures,patterns of embryonic development,
and common ancestry.

14 points—Answer a maximum of three.

QUESTION #4

Scientists seeking to determine which molecule is responsible for the transmission of characteristics from one generation to the next knew that the molecule must (1) copy itself precisely, (2) be stable but able to be changed, and (3) be complex enough to determine the organism's phenotype.

Explain how DNA meets each of the three criteria stated above
Select **one** of the criteria stated above and describe experimental evidence used to determine that DNA is the hereditary material. One point for each response.

Note: Part A can earn a maximum of 8 points. Part B can earn a maximum of 2 points.

Explain how DNA meets each of the three criteria stated above
[MAX 8]

1. Molecular Properties for Precise COPYING **[MAX 3]**

   (A) Template concept (semi-conservative replication)

   (B) Molecular structure [e.g., complementary base pairing; A:T,C:G; purine-pyrimidine pairing; antiparallel]

   (C) DNA polymerase function in copying

   (D) Separation concept

   (E) Elaboration of replication [e.g., specific roles of other replication enzymes, proper sequence of steps]

2. Molecular Properties that make it STABLE but ABLE to change [MAX 3]

Stable

(A) Energetically favorable arrangement; stable because of shape of molecule [e.g., double helix, bases in the interior of the helix]

(B) Energetically favorable arrangement; stable because of bonding [e.g., multiple H bonds, phosphodiester bonds]

(C) Silent errors [e.g., "junk" DNA, introns, redundancy of the genetic code]

(D) Able to be repaired [e.g., proofreading]

Able

(A) Description of a mutation [e.g., substitution, deletion, insertion, inversion, translocation]

(B) Crossing over [e.g., during meiosis]

(C) Base changes [e.g., depurination, deamination, tautomerism]

(D) Gene rearrangements [e.g., antibody genes in stem cells, transposons]

(E) Sensitive to mutagens [e.g., UV, X-ray]

(F) Restriction enzyme recognition sequences

Stable or Able (with justification)

(A) Methylation

(B) Telomeres protect ends

3. Molecular Properties that make it COMPLEX enough to determine PHENOTYPE [**MAX 3**]

   (A) Colinearity of gene and protein [i.e., base sequence determines *aa* sequence]

   (B) Infinite base sequence combinations lead to protein variety

   (C) Variable numbers of base pairs per gene lead to different sizes of polypeptides

   (D) Proteins are responsible for phenotype

   (E) Description of transcription and translation

   (F) Chromosome structure as it relates to function [e.g., supercoiling, chromosome or gene inactivation, interaction with histones, etc.]

Select **one** of the criteria stated above and describe experimental evidence used to determine that DNA is the hereditary material [**MAX 2**]

   Identification of experiment with valid link to any one of the criteria [**MAX 1**]:

1. precise copying

2. stable but able to change

3. complex enough to determine phenotype

4. DNA is the hereditary material

May include but need not be limited to experiments that show:

1. DNA can transform bacteria

2. viral DNA can reprogram cells

3. equivalence of A:T and C:G

4. double helix structure [e.g., x-ray crystallography]

5. replication is semiconservative

6. hereditary enzyme deficiency disorders have genetic links

7. DNA codes for protein

8. changes in DNA quantities during the cell cycle (mitosis/meiosis)

9. chromosome markers linked to disorders

10. measurement of mutation rates

11. changes in DNA [biotechnology]

Description of EVIDENCE [**MAX 1**]
Evidence provided in the context of an experiment

PRACTICE TEST 1

AP BIOLOGY

SECTION I

Time–1 Hour and 30 Minutes
Number of Questions–120
Percent of Total Grade–60

Directions: Each of the questions or incomplete statements below is followed by five suggested answers. Select the one that is best in each case.

1. Which of the following best explains surface tension?

   (A) Hydrogens of one water molecule are attracted to the hydrogens of another water molecule.

   (B) Oxygen of one molecule is attracted to the hydrogens of the same molecule.

   (C) Hydrogen of one water molecule is attracted to the oxygen of another water molecule.

   (D) Water molecules increase the buoyancy of each other.

   (E) Water molecules have a high heat of evaporation.

2. Which of the following bond types is the strongest?

   (A) ionic

   (B) covalent

   (C) van der Waals

   (D) hydrophobic

   (E) hydrogen

3. Consider the following chemical formula:

   NH_2-CH_2-COOH

   The last part of this molecule, which is underlined, is said to be a(n)

   (A) carboxyl group

   (B) carbon backbone

   (C) amine group

   (D) alcohol

   (E) aldehyde

4. The molecule in question 3 is best described as a(n)

   (A) ketone

   (B) alcohol

   (C) benzoic acid

   (D) fatty acid

   (E) amino acid

5. The tertiary structure of proteins is due to

(A) the sequence of amino acids in the polypeptide

(B) the bonding between side chains or variable groups

(C) the regular spacing of hydrogen bonds

(D) the interaction of subunits or large polypeptide chains

(E) none of the above

6. The diagram below represents Watson and Crick's model of DNA.

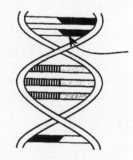

The substance indicated by the arrow could be

(A) thymine

(B) deoxyribose

(C) ribose

(D) uracil

(E) phosphate

7. A student was given 10 ml of a 10M HCl stock solution and was asked to prepare 2 ml of a 0.2M solution. How much of the stock solution does she need to use?

(A) 0.1 ml

(B) 1.0 ml

(C) 0.2 ml

(D) 0.04 ml

(E) 0.02 ml

8. The enzyme ATP synthase is responsible for making ATP and is found in large quantities embedded in mitochondrial and chloroplast membranes. Peter Mitchell proposed the idea of chemiosmosis to help explain ATP synthesis. Choose the answer that best completes the following sentence: Hydrogen ions are driven from the mitochondrial _____ to the inner membrane space by the power of the _____ and then flow down a gradient through ATP synthase, phophorylating ADP to produce ATP.

(A) cristae; ATP

(B) matrix; ATP

(C) cristae; electron transport chain

(D) matrix; electron transport chain

(E) matrix; cristae

9. The following graph needs to be labeled by a forgetful student who was graphing the reaction rate of the decomposition of hydrogen peroxide by catalase. How does he label the X (independent) axis and the Y (dependent) axis, respectively?

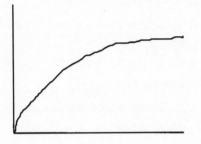

(A) rate of reaction; enzyme concentration

(B) rate of reaction; substrate concentration

(C) enzyme concentration; substrate concentration

(D) enzyme concentration; rate of reaction

(E) substrate concentration; rate of reaction

10. One of the major differences between prokaryotic and eukaryotic cells is the size of their

(A) nuclei

(B) ribosomes

(C) cytoskeleton

(D) endoplasmic reticulum

(E) none of the above

11. Organelles that are not part of the endomembrane system, and therefore do not take part in cellular secretions, include which of the following?

(A) smooth endoplasmic reticulum

(B) rough endoplasmic reticulum

(C) peroxisomes

(D) Golgi apparatus

(E) lysosomes

12. Which of the following organelles is not matched with its corresponding function?

(A) ribosome: protein synthesis

(B) nucleolus: ribosome production

(C) Golgi apparatus: secretion of products

(D) microtubules: muscular contraction

(E) lysosomes: digestion

13. Which one of the following substances is most likely to pass through a selectively permeable plasma membrane?

(A) O_2

(B) K_+

(C) glucose

(D) NH_3

(E) starch

14. Which one of the following statements about cell division is false?

(A) There are regulatory proteins called cyclins that phosphorylate enzymes that begin the cell cycle.

(B) Cyclins build up after cell division.

(C) Cdks or cyclin dependent kinases are present in uniform concentrations throughout the cell cycle.

(D) Cdks combine with cyclins during the cell cycle.

(E) Cell division is influenced by both growth factors and cell density.

15. Which of the following statements is true?

    (A) Colchicine is a mitotic poison because it interferes with growth factors.

    (B) Colchicine is a mitotic poison because it destroys centrioles.

    (C) Colchicine is a mitotic poison because it destroys microtubules.

    (D) Colchicine application to a growing leaf would leave the number of chromosomes the same in a cell after division.

    (E) Colchicine destroys the ends of chromosomes, known as telomeres, and therefore causes cells to cease dividing and die.

16. The phases below describe several events that occur during the process of mitosis.

    (1) Attachment of double-stranded chromosomes to the spindle apparatus

    (2) Formation of single-stranded chromosomes, which are moved to opposite ends of the cell

    (3) Disintegration of the nuclear membrane

    (4) Nuclear membrane formation around each set of chromosomes, forming two nuclei

    (5) Synthesis of a spindle apparatus

    Which sequence represents the correct order of these events?

    (A) $1 \rightarrow 2 \rightarrow 3 \rightarrow 4 \rightarrow 5$

    (B) $2 \rightarrow 4 \rightarrow 1 \rightarrow 3 \rightarrow 5$

    (C) $1 \rightarrow 4 \rightarrow 5 \rightarrow 3 \rightarrow 4$

    (D) $3 \rightarrow 5 \rightarrow 1 \rightarrow 2 \rightarrow 4$

    (E) $2 \rightarrow 4 \rightarrow 1 \rightarrow 5 \rightarrow 3$

17. The diagram below represents the fluid mosaic model of a plasma membrane. What do letters X and Y indicate?

    (A) X indicates pores and Y indicates cellulose

    (B) X indicates phospholipids and Y represents proteins

    (C) X indicates proteins and Y indicates phospholipids

    (D) X indicates nuclei and Y indicates ribosomes

    (E) X indicates lipids and Y represents phospholipids

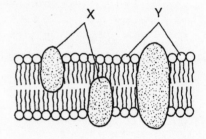

18. Which of the following does not use mitosis?

    (A) wound healing

    (B) development from blastula to gastrula

    (C) bone growth

    (D) liver regeneration

    (E) production of sex cells

19. During mitosis, chromosomes are not visible as discrete units during

    (A) prophase

    (B) metaphase

    (C) anaphase

    (D) interphase

    (E) telophase

20. Which of the following statements about enzymes is **NOT** true?

(A) Competitive inhibitors act away from the active site.

(B) Allosteric inhibitors act away from the active site.

(C) Allosteric inhibitors can change the size of the active site.

(D) Competitive inhibitors usually resemble the substrate.

(E) Noncompetitive inhibitors are allosteric inhibitors.

21. A researcher isolated mitochondria and placed them into a buffered pH 4 solution. After several hours, he replaced the solution with a buffered pH 8 solution. A test revealed more ATP 30 minutes after transfer to the second solution than there was after placing them in the first solution. The most likely explanation for this is

(A) Mitochondria produce more ATP sythase at higher pHs.

(B) Hydrogen ions were flowing through ATP synthase after the second transfer.

(C) Mitochondria have an optimum pH of 8.

(D) ATP synthase has a competitive inhibitor whose optimum pH is 8.

(E) The electron transport chain is more active at pH 8.

22. Most of the ATP made during cellular respiration is generated by

(A) glycolysis

(B) intermediate transport steps

(C) the Krebs cycle

(D) oxidative phosphorylation

(E) photophosphorylation

23. The product(s) of alcohol fermentation is (are)

(A) lactic acid

(B) ethanol

(C) ADP

(D) NADH

(E) both ADP and NADH

24. At the end of glycolysis, each molecule of glucose has yielded 2 molecules of _____, 2 molecules of _____, and a net of 2 molecules of _____.

(A) lactic acid; NADH; ATP

(B) ethanol; NAD^+; ATP

(C) pyruvate; NADH; ADP

(D) pyruvate; NAD^+; ADP

(E) pyruvate; NADH; ATP

25. In anaerobic cells, the ratio of pyruvate/lactate is much less than 1 while under aerobic conditions, the ratio of pyruvate/lactate is much greater than 1 because

(A) pyruvate is the oxidizing agent in the formation of NAD^+ from NADH

(B) lactic acid can only be generated under anaerobic conditions required for ADP regeneration

(C) pyruvate decomposes to lactate under anaerobic conditions

(D) pyruvate is converted to carbon dioxide under anaerobic conditions

(E) mitochondria prefer lactic acid to pyruvate for cellular respiration pathways

26. Oxygen is required for aerobic cellular respiration because it is

    (A) an activator for pyruvate kinase

    (B) the reducing agent in the electron transport chain

    (C) the oxidizing agent in the electron transport chain

    (D) the oxidizing agent for acetyl CoA

    (E) the reducing agent for acetyl CoA

27. Which of the following statements is true about phosphofructokinase?

    (A) It is stimulated by ATP.

    (B) It is stimulated by citrate.

    (C) Acetyl CoA is one of its substrates.

    (D) Fructose phosphate is one of its products.

    (E) Fructose bisphophate is one of its products.

28. The products of the light reactions in photosynthesis are

    (A) oxygen and $NADP^+$

    (B) water and NADPH

    (C) oxygen and NADPH

    (D) water and oxygen

    (E) oxygen and NAD^+

29. Which of the following statements is true about RuBP (ribulose bisphosphate) ?

    (A) It is the 3 carbon product of the Calvin cycle.

    (B) It is the final oxidizing agent in the light reaction.

    (C) It is the CO_2 acceptor in the Calvin cycle.

    (D) It is the rarest substate in the Calvin cycle and therefore the limiting reagent.

    (E) None of the above are true.

30. If 6 molecules of oxygen are released during photosynthesis, how many molecules of carbon dioxide would be fixed?

    (A) 1

    (B) 3

    (C) 6

    (D) 12

    (E) 24

Questions 31 and 32 refer to the karyotype below.

31. Information in this karyotype indicates that the individual is a

    (A) female with sickle cell anemia

    (B) male with Tay-Sachs disease

    (C) female with Down syndrome

    (D) male with Down syndrome

    (E) female with Tay-Sachs disease

32. The number of pairs of homologous chromosomes in the karyotype above is

    (A) 0

    (B) 21

    (C) 23

    (D) 25

    (E) 46

33. The number of chromosomes in normal humans is 46. In sperm production, the number of chromosomes after meiosis I would be_____, and the number of resulting cells would be_____.

(A) 46; 2

(B) 23; 2

(C) 46; 1

(D) 23; 1

(E) 23; 4

34. Variation within a population can arise through which of the following means?

(A) Nondisjunction only

(B) Polyploidy only

(C) Crossing over only

(D) Nondisjunction and polyploidy

(E) Nondisjunction, polyploidy, and crossing over

35. Use the drawing below to answer the following question.

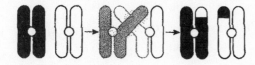

Which process is illustrated by the diagram above?

(A) nondisjunction

(B) crossing-over

(C) sex determination

(D) independent assortment

(E) none of the above

36. The diagram below represents four beakers, each containing an equal number of two colors of beads.

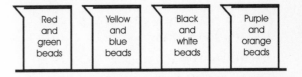

One bead was removed at random from each of the four beakers, and the colors were recorded. The beads were then returned to the original beakers. When the procedure was repeated several times, different combinations of colored beads were obtained. This activity could best be used to illustrate

(A) mitotic cell division

(B) sex linkage

(C) crossing-over

(D) independent assortment

(E) none of the above

37. In gray squirrels, gray body color is dominant to red, and straight hair is dominant to curly. A male gray squirrel with straight hair and a red female with straight hair had 8 pups:

3 gray straight, 3 red straight, 1 gray curly, and 1 red curly.
The genotypes of the parents were

(A) GgSs [X] GgSs

(B) GgSs [X] ggSs

(C) Ggss [X] GgSs

(D) ggSs [X] ggSs

(E) Ggss [X] Ggss

38. Two parents with which two blood phenotypes could produce a child with type O blood?

(A) A [X] B

(B) AB [X] O

(C) AB [X] A

(D) AB [X] AB

(E) none of the above

39. A male with normal color vision marries a woman whose father was red-green color-blind but who herself can see colors normally. What is the chance their first child will be a color-blind boy? (Red-green color-blindness is a sex-linked trait.)

(A) She cannot have a color-blind boy.

(B) All her male children will be color-blind.

(C) 1 in 2

(D) 1 in 4

(E) 1 in 8

40. In peas, green color is dominant to yellow. All peas will be either one or the other color so long as they do not have the dominant form of the colorless gene C. The allele C prevents the deposition of pigments in the seeds, while c has no effect. If I crossed two plants that were both heterozygous for both traits, the ratio of the offspring would be

(A) 9/16 colorless: 3/16 green : 4/16 yellow

(B) 9/16 green: 3/16 yellow: 4/16 colorless

(C) 1/2 green: 1/2 yellow

(D) 12/16 colorless: 3/16 green: 1/16 yellow

(E) there is not enough information given to answer the question

41. The situation above is typical of

(A) crossing-over

(B) epistasis

(C) nondisjunction

(D) linked genes

(E) sex linkage

42. Watson and Crick used all of the following information in elucidating the physical structure of DNA except for

(A) X-ray crystallography

(B) the Meselson-Stahl experiment

(C) Chargoff's rules

(D) the different sizes of purines and pyrimidines

(E) the transforming principle of Avery *et al*

43. In an important experiment in the early fifties that helped prove DNA was the hereditary material, Hershey and Chase found that

(A) the radioactive phosphorus they used was bound in bacterial DNA

(B) the radioactive sulfur they used was bound in bacterial DNA

(C) the radioactive phosphorus they used was bound in viral DNA

(D) the radioactive sulfur they used was bound in viral DNA

(E) two thirds of the phosphorus was bound to the viral coat

44. DNA can act like a template for RNA because they

(A) are both single stranded

(B) can match sugars and phosphates

(C) have the same nitrogen bases

(D) both use histones for organization

(E) can use base pairing

45. The number of times that you would expect the restriction enzyme BAM HI, which cuts the sequence CATTAC, to cut the 3 million base pair sequence of *E. coli* is about

(A) 2

(B) 20

(C) 70

(D) 700

(E) 2000

Questions 46–48 refer to the diagram below.

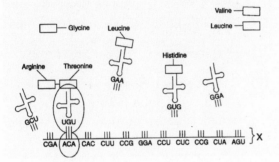

46. The synthesis of structure X occurred in the

(A) nucleus

(B) cytoplasm

(C) lysosome

(D) vacuole

(E) endoplasmic reticulum

47. The amino acid that would be transferred to the position of codon CAC would be

(A) leucine

(B) histidine

(C) valine

(D) glycine

(E) arginine

48. The biochemical process represented in the diagram is most closely associated with the cell organelle known as the

(A) nucleolus

(B) chloroplast

(C) ribosome

(D) mitochondrion

(E) medichlorion

49. The minimum number of different types of transfer RNAs (tRNAs) that a person can have is

(A) 4

(B) 16

(C) 20

(D) 45

(E) 88

50. Which of the following can influence the amount of gene expression in prokayotes?

(A) Introns

(B) Enhancers

(C) Poly-A tail

(D) 5 prime guanine cap

(E) None of the above

51. All of our cells contain proto-oncogenes, which may turn into oncogenes, which are cancer genes. The best explanation for proto-oncogenes is that they

(A) came into our cells from a viral infection of our ancestors

(B) arose from plasmids that have been inserted into bacteria and now reside in us

(C) are DNA "junk" with no known function

(D) turn into oncogenes as we age

(E) help regulate cell division

52. In genetic engineering, it is necessary to cut DNA of bacteria and eukaryotes. To insert the human insulin gene into a bacteria, one should use

(A) two different restriction enzymes so that the pieces won't reanneal

(B) the same restriction enzyme so that both pieces will have the same sticky ends

(C) methylated bacterial DNA so that only some of it will be spliced

(D) two different ligases to reanneal the DNA

(E) a hot water bath at 55 degrees centigrade so that the reaction will happen faster

53. It is assumed that in Earth's early history

(A) autotrophs were rare

(B) heterotrophs were rare

(C) nitrogen was rare

(D) prokaryotes were rare

(E) oxygen was plentiful

54. Darwin's finches are a group of sparrow-like birds that inhabit the Galapagos Islands. Which of the following statements about them is most likely true?

(A) Darwin's finches are good examples of convergent evolution because they are closely related.

(B) Darwin's finches are good examples of convergent evolution because they look so similiar.

(C) Darwin's finches are good examples of adaptive radiation because they come from a relatively recent ancestor.

(D) Darwin's finches have had 100 million years to become so different in appearance.

(E) Darwin's finches are good examples of adaptive radiation because of the beneficial radiation they receive living close to the equator.

55. Homologies are good evidences of evolution. All of the following are homologies except the

(A) forearms of bats and humans

(B) wings of bats and birds

(C) DNA of rabbits and birds

(D) cytochrome-C of bats and birds

(E) dorsal nerve cord of whales and bats

56. The changes in foot structure in a bird population over many generations are shown in the diagram below.

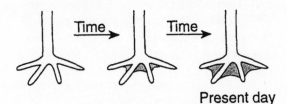

Present day

These changes can best be explained by the concept of

(A) evolution

(B) extinction

(C) stable gene frequencies

(D) use and disuse

(E) Lamarkism

57. One way that the modern theory of evolution differs from the theory of evolution proposed by Charles Darwin is that the modern theory

(A) includes the concept of use and disuse

(B) capitalizes on the geological work of Charles Lyell

(C) does not include the concept of overproduction

(D) identifies mutations as one explanation for variations

(E) does not consider the effects of genetic change

58. Using the graphs below, identify the one that best represents the pattern of evolution according to the concept of punctuated equilibrium.

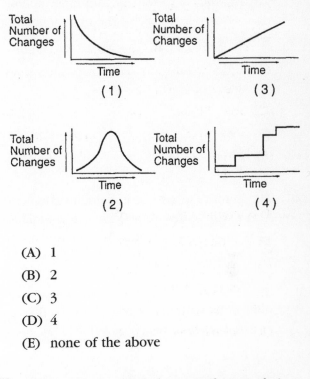

(A) 1

(B) 2

(C) 3

(D) 4

(E) none of the above

59. What will most likely happen if a population is large and no migration, mutation, or environmental change occurs?

(A) Natural selection will increase.

(B) Nonrandom mating will start to occur.

(C) The rate of evolution will increase.

(D) Genetic drift will increase.

(E) Gene frequencies will remain constant.

60. Four-o-clocks are flowers that exhibit incomplete dominance in their petal color. There are red and white homozygous flowers, but the heterozygote is pink. If one planted 6 red and 4 white four-o-clocks in a garden, and there was random cross pollination, what percentage of the next generation would be pink?

(A) 100%

(B) 0%

(C) 36%

(D) 48%

(E) 52%

61. The number of people in certain regions of Africa who suffer from sickle cell anemia is 14 percent. This genetic disorder is caused by a mutation that when homozygous recessive results in sickle cell anemia, but in the heterozygote condition causes sickle cell trait. What is the gene frequency for this mutated allele in that population?

(A) 0.14

(B) 0.37

(C) 0.63

(D) 0.86

(E) there is not enough information given to answer the question

62. Speciation is the formation of new species. The most common cause of speciation is

(A) mutation

(B) natural selection

(C) geographical isolation

(D) recombination

(E) gene flow

63. In 1889, August Weismann conducted an experiment attempting to produce mice without tails. He cut the tails off adult mice and then permitted them to mate. All offspring had long tails. He repeated the experiment many times, for many generations, always with the same results. This experiment helped to disprove the concept of

(A) overproduction in species

(B) inheritance of acquired characteristics

(C) survival of the fittest

(D) struggle for existence

(E) evolution

64. Viruses are considered an exception to the cell theory because they

(A) normally reproduce by binary fission

(B) normally reproduce only in a host cell

(C) normally reproduce by forming spores

(D) lack DNA or RNA

(E) have no cell organelles

65. The diagram below represents two cells next to a metric measuring device under the low-power objective of a compound light microscope.

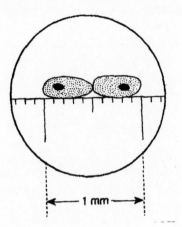

What is the approximate length of a nucleus of one of these cells?

(A) 50μm

(B) 100μm

(C) 500μm

(D) 1000μm

(E) 1500μm

66. Prokaryotes are divided into two major groups, the Eubacteria and the Archaebacteria, two domains suggested by Carl Woese and others. Which of the following statements about the two domains is **NOT** true?

(A) They diverged very early in evolutionary history.

(B) They both have peptidoglycan in their cell walls.

(C) Of the two groups, the Archaebacteria are most closely related to plants and animals.

(D) Eubacteria include the cyanobacteria.

(E) Archaebacteria have membrane enclosed organelles; Eubacteria do not.

67. Because mushrooms absorb nutrients from decaying leaves, they are classified as

(A) autotrophs

(B) anaerobes

(C) protozoans

(D) heterotrophs

(E) parasites

68. Mitochondria exist in

(A) all eukaryotes

(B) all archaebacteria

(C) all eubacteria

(D) all archaebacteria and eukaryotes

(E) all of the above

69. The flowing cytoplasm of an active amoeba is an adaptation for

(A) extracellular digestion

(B) glucose synthesis

(C) intracellular circulation

(D) lipid synthesis

(E) nerve transmission

Questions 70 and 71 refer to the diagram below.

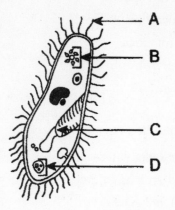

70. The diagram represents a freshwater protist. Which letter indicates a structure that prevents the accumulation of excess water in the protist?

(A) A only

(B) B only

(C) C only

(D) D only

(E) A, B, C, and D

71. The organism depicted above is a

(A) ciliate and a protozoan

(B) flagellate and a protozoan

(C) sporozoan and a protozoan

(D) sporozoan and a prokaryote

(E) ciliate and a prokaryote

72. Choose the correct sequence of taxonomic terms from the most to least inclusive in the lists that follow.

(A) class, phylum, order, family, genus

(B) phylum, kingdom, class, order, family

(C) kingdom, phylum, order, class, family

(D) domain, kingdom, phylum, order, class

(E) kingdom, class, order, family, genus

73. Scientists found several marine organisms heretofore unknown or unclassified. The organisms were radially symetrical and diploblastic, that is, they lacked a mesoderm. To what existing phylum should they be assigned?

(A) Porifera

(B) Cnidaria

(C) Platyhelminthes

(D) Rotifera

(E) Echinodermata

74. Organisms that possess a body cavity not enclosed by mesoderm are termed pseudocoelomates. Which of the following groups are pseudocoelomates?

(A) Echinodermata

(B) Mollusca

(C) Arthropoda

(D) Annelida

(E) Nematoda

75. Organisms that possess true coeloms may be divided into two groups, the protostomes and the deuterostomes. Which of the following characteristics would place an animal in the protostome group?

(A) Determinate cleavage

(B) Spiral cleavage

(C) Blastophore forms the mouth

(D) Coelom forms from splits in the meso-derm

(E) All of the above

76. Vascular plants are grouped according to whether they have seeds or not. Which of the following produce seeds?

(A) Club mosses

(B) Horsetails

(C) Conifers

(D) All of the above

(E) None of the above

77. In chordates, locomotion is accomplished by the action of

(A) jointed chitinous appendages attached to muscles

(B) paired setae attached to muscles

(C) muscles attached to an exoskeleton

(D) paired muscles attached to an endoskeleton

(E) muscles attached to a hydrostatic skeleton

78. Which structures are directly involved in the process of excretion in an earthworm?

(A) nephridia and moist skin

(B) Malpighian tubules and nephrons

(C) tracheal tubes and moist skin

(D) nephridia and spiracles

(E) Malpighian tubules and spiracles

79. In the grasshopper and the earthworm, the efficiency of food absorption is increased by the presence of

(A) a large stomach, which secretes hormones

(B) infolds, which add surface area to the digestive tube

(C) the pancreas, which secretes digestive enzymes into the intestine

(D) the liver, which absorbs excess sugar and stores it as glycogen

(E) the cecum, which stores unsuitable intake for excretion

80. An animal maintains its fluid balance by regulating the gain and loss of water. This maintenance is an example of

(A) excretion

(B) hydrolysis

(C) homeostasis

(D) cyclosis

(E) peristalsis

81. Which statement is **NOT** true of ferns?

(A) The gametophyte is larger than the sporophyte.

(B) The sori give rise to gametophytes.

(C) Eggs are fertilized by sperm in the archegonia.

(D) Ferns possess vascular tissue for the transport of water and minerals.

(E) Male gametes depend on water for fertilization.

82. Two major groups within flowering plants (Anthophyta) are monocotyledons and dicotyledons. Which of the following is **NOT** true of dicotyledons?

(A) They possess floral parts in multiples of fours or fives.

(B) They have two seed leaves, not one.

(C) Their vascular bundles are not scattered randomly in the stem but are aligned in a ring fashion.

(D) Their leaf venation is netted, not parallel.

(E) They have a fibrous root system.

83. Which of the following statements is **NOT** true concerning the cohesion-tension theory of water movement in plants?

(A) Adhesion accounts for the attraction between water molecules.

(B) Cohesion accounts for the attraction between water and the vessel walls.

(C) Root pressure in a significant force in driving water up a stem.

(D) Companion cells help xylem regulate the passage of water and minerals.

(E) All of the above are not true.

84. Which of the following statements is true concerning plants that utilize C4 photosynthesis?

(A) They only open their stomates at night.

(B) They use bundle sheath cells to separate the Calvin cycle from the atmosphere.

(C) They use rubisco (ribulose bisphosphate carboxylase oxygenase) to capture the carbon dioxide from the atmosphere.

(D) There are more C4 plants in temperate deciduous forests because of the mild heat and available moisture.

(E) All of the above statements are true.

85. Neurotransmitters are chemical substances released by cells that act on other cells. They could affect

(A) nerve cells

(B) endocrine cells

(C) muscle cells

(D) all of the above

(E) none of the above

86. Neurotransmitters cross the synapse after being released from the

(A) dendrites

(B) cell bodies

(C) axons

(D) nuclei

(E) Nodes of Ranvier

87. Insulin is an important hormone that serves to decrease one's blood sugar. Glucagon has the opposite effect through

(A) the conversion of glycogen to glucose

(B) the secretion of glucose by cells

(C) causing gluconeogenesis

(D) all of the above

(E) none of the above

88. The target tissue(s) for parathyroid hormone is (are)

(A) the kidneys

(B) the bones

(C) the small intestines

(D) all of the above

(E) none of the above

89. Oxytocin is produced in the

(A) muscles

(B) pancreas

(C) thyroid

(D) all of the above

(E) none of the above

Questions 90–92 refer to the following diagram.

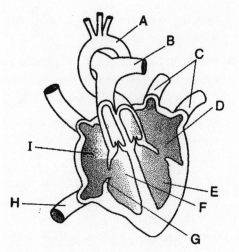

90. Which structures are most closely associated with the transport of deoxygenated blood?

(A) A, B, and C

(B) B, F, and I

(C) C, D, and E

(D) D, H, and I

(E) A, B, and I

91. A structure that prevents backflow into an atrium is indicated by letter

(A) G

(B) B

(C) C

(D) H

(E) E

92. The sinoatrial node or pacemaker is located closest to the region indicated by letter

(A) A

(B) I

(C) F

(D) D

(E) E

93. A similarity between the nervous system and the endocrine system in humans is that they both

(A) are composed of neurons

(B) are composed of glands

(C) maintain homeostasis

(D) secrete chemicals across synapses

(E) are composed of myofibrils

Questions 94-96 refer to the following diagram.

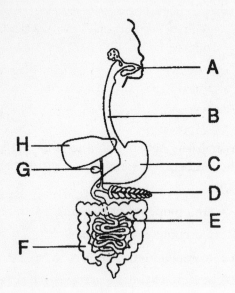

94. In which structure does the initial hydrolysis of carbohydrates occur?

(A) A

(B) H

(C) C

(D) F

(E) E

95. From which structure are glucose and amino acids normally absorbed into the circulatory system?

(A) A

(B) H

(C) C

(D) F

(E) E

96. In which structure does extracellular chemical digestion of protein begin?

(A) A

(B) H

(C) C

(D) D

(E) E

97. Which is true of the hormones involved in digestion?

(A) Gastrin is secreted by the stomach and acts on the stomach.

(B) Cholecystokinin (CCK) is produced by the small intestine and acts on the gall bladder, causing it to release bile.

(C) Secretin is produced by the intestine and causes the pancreas to release bicarbonate.

(D) Enterogastrone is secreted by the small intestine and acts on the stomach to reduce peristalsis.

(E) All of the above are true.

246

Questions 98 and 99 refer to the following diagrams.

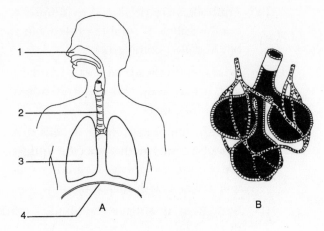

98. Diagram B represents the functional unit of which structure represented in diagram A?

(A) 1

(B) 2

(C) 3

(D) 4

(E) all of the above

99. Choose the false statement from among the following choices.

(A) Number 2 represents the esophagus and divides into bronchi.

(B) Number 2 represents the trachea and has cilia lining it.

(C) Number 4 represents the diaphragm, which by contraction lowers the pressure in the lungs.

(D) Number 3 contains many tubes called bronchioles.

(E) All of the above statements are false.

100. Which sequence represents the correct pathway for the removal of urine from the human body?

(A) kidney → urethra → urinary bladder → ureter

(B) ureter → kidney → urinary bladder → urethra

(C) urethra → kidney → urinary bladder → ureter

(D) kidney → ureter → urinary bladder → urethra

(E) kidney → urinary bladder → ureter → urethra

Questions 101 and 102 refer to the following drawing.

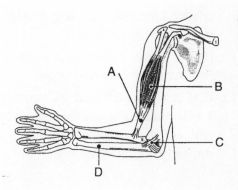

101. A ligament is represented by the structure labeled

(A) A

(B) B

(C) C

(D) D

(E) none of the above

102. The forearm bone, represented by the letter D, and the bone proximal to it are respectively the

(A) humerus and radius

(B) humerus and ulna

(C) radius and metacarpel

(D) ulna and humerus

(E) ulna and metacarpel

103. Which of the following statements about CD4 or helper T cells is false?

(A) Helper T cells use cytokines to stimulate both cytotoxic T cells and B cells.

(B) Helper T cells receive their stimulus from the combination of class I MHC protein and antigen.

(C) Helper T cells are the cells most infected by the HIV virus.

(D) Helper T cells cannot respond to free floating antigens in the blood.

(E) CD4 is a protein found on most helper T cells that facilitates the interaction between the class II MHC protein and the antigen producing cell.

104. Which of the following statements about antibodies is false?

(A) Antibodies are produced by different combinations of soluble proteins in the blood called complement.

(B) When antibodies bind to a bacterial cell, it may keep that cell from adhering to host cells.

(C) The specific receptors on T cells that would interact with a specific antigen are the same shape as antibodies that will interact with that antigen.

(D) By binding to antigens, antibodies mark those antigens for destruction by macrophages.

(E) By binding to an invader, antibodies attract complement proteins, which lyse the target membrane.

Questions 105–106 refer to the drawing below.

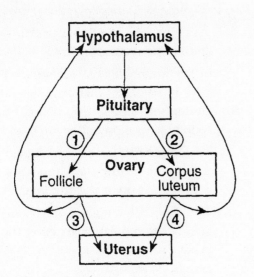

105. Which hormone(s) maintain(s) the uterine lining after ovulation?

(A) 1 only

(B) 2 only

(C) 4 only

(D) 3 and 4

(E) 2 and 4

106. Which hormone(s) most directly promote(s) the maturation of the egg?

(A) 1 only

(B) 2 only

(C) 3 only

(D) 1 and 2

(E) 2 and 4

107. Which of the following statements about the menstrual cycle is **FALSE**?

(A) Luteinizing hormone is produced by the pituitary in response to GnRH from the hypothalamus.

(B) Luteinizing hormone acts on the follicle, causing ovulation, around the fourteenth day of the cycle.

(C) Birth control pills that contain estrogens and progesterones would inhibit the secretion of FSH by the pituitary.

(D) The phase of menstruation at which pregnancy is most likely to occur is the secretory phase.

(E) FSH and estrogen production are highest during days 25–28 of the menstrual cycle.

108. Which of the following is the correct sequence of events depicting a nerve stimulating a voluntary muscle?

(A) Neurotransmitter release → calcium ion release → sodium influx → troponin binding → tropomyosin twisting → myosin and actin binding → ATP utilization

(B) Neurotransmitter release → sodium influx → calcium ion release → troponin binding → tropomyosin twisting → myosin and actin binding → ATP utilization

(C) Neurotransmitter release → calcium ion release → sodium influx → troponin binding → tropomyosin twisting → ATP utilization → myosin and actin binding

(D) Neurotransmitter release → sodium influx → calcium ion release → tropomyosin twisting → troponin binding → ATP utilization → myosin and actin binding

(E) Neurotransmitter release → sodium influx → troponin binding → tropomyosin twisting → myosin and actin binding → Calcium ion release → ATP utilization

109. Voltage-gated channels help create and regulate action potentials or nerve impulses. Choose the correct statement from the following choices.

(A) During the resting state, the sodium activation and inactivation gates are both open.

(B) During the depolarizing phase, the sodium activation and inactivation gate are both open.

(C) During the resting state, the sodium activation is open and the inactivation gate is closed.

(D) During the repolarizing phase, the voltage-sensitive potassium activation gate is closed.

(E) During the undershoot, all three types of gates are closed.

Questions 110–111 refer to the following diagram.

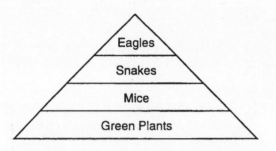

110. In the food pyramid represented above, which level has the greatest biomass?

(A) Eagles

(B) Snakes

(C) Mice

(D) Green plants

(E) There is no way to tell from the amount of information given.

111. If 6 million kilocalories per year per square meter fell on the environment where the green plants are growing, the number of kilocalories per year per square meter bound up in the snakes would be approximately

(A) 120,000

(B) 60,000

(C) 6,000

(D) 600

(E) 60

112. Several years after a building had been torn down and the ground cleared, grasses began to grow in that area. After ten years, small bushes replaced the grasses. This pattern of plant growth is known as

(A) biological control

(B) ecological succession

(C) land-use management

(D) cover cropping

(E) biological magnification

113. Symbiotic relationships frequently develop between organisms that live in close physical proximity to each other. Female cowbirds often lay their eggs in the nests of wrens, who unwittingly raise the young cowbirds as their own. This relationship is an example of

(A) commensalism

(B) parasitism

(C) mutualism

(D) competition

(E) nihilism

114. The diagram below shows a relationship between altitude, latitude, and tree growth.

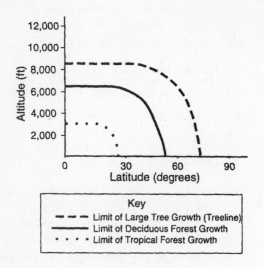

What is a valid inference that can be made based on this graph?

(A) Deciduous trees cannot grow at an altitude of 5,000 feet.

(B) The effects of increasing altitude and latitude on tree growth are similar.

(C) There is less light available at 10,000 feet than at lower elevations.

(D) Trees do not grow rapidly in the tropics.

(E) Deciduous trees grow faster than tropical trees.

115. The graph below shows the changes in two populations of herbivores in a grassy field.

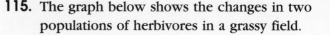
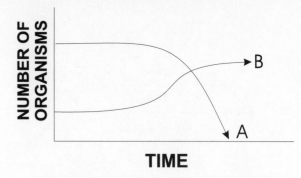

A possible reason for these changes is that

(A) all of the plant populations in this habitat decreased

(B) population A produced more offspring than population B did

(C) population A consumed the members of population B

(D) members of population B were larger than population A

(E) population B competed more successfully for food than population A did

116. A fungus living on a dead tree is a

(A) saprophyte

(B) heterotroph

(C) consumer

(D) all of the above

(E) none of the above

Questions 117–119 refer to the following diagram, which shows the general location of some major biomes of the Western Hemisphere.

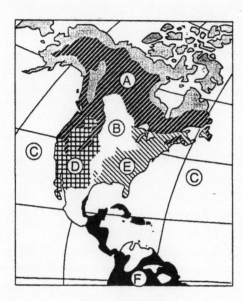

117. The most stable biome is represented by letter

(A) A

(B) B

(C) C

(D) D

(E) E

118. Which type of biome is indicated by letter E?

(A) Tundra

(B) Taiga

(C) Grassland

(D) Deciduous forest

(E) Chaparral

119. The biome indicated by letter F is characterized by

(A) heavy rainfall, broad leaved plants, and monkeys

(B) sparse rainfall, grasses, and leopards

(C) variable rainfall, mosses, and caribou

(D) heavy rainfall, conifers, and snakes

(E) variable rainfall, moderate temperatures, and low humidity

120. The return of salmon to their own birthplace to breed is an example of

(A) habituation

(B) a stimulus

(C) operant conditioning

(D) reasoned behavior

(E) imprinting

SECTION II

Time–1 Hour and 30 Minutes
Number of Questions–4

Directions: Answer all questions. Number your answer as the question is numbered below. Answers must be in essay form. Outline form is NOT acceptable. Labeled diagrams may be used to supplement discussion, but in no case will a diagram alone suffice. It is important that you read each question completely before you begin to write.

1. Proteins are essential for almost all aspects of life. By using location, structure, and function, describe the important proteins in two of the following three systems:

    (A) cell cycle and division

    (B) cell movement

    (C) signal transduction and communication

2. Since Watson and Crick's elucidation of the structure of DNA, many scientists have sought to concentrate their research efforts on DNA, the code for life. Taxonomists and phylogeneticists use changes in DNA to study classification and evolution. Describe

    (A) all types of DNA within prokaryotic and eukaryotic cells

    (B) the sources of variability in DNA sequences and chromosome number

    (C) how scientists have studied changes or differences in DNA, using the theme of science as a process

3. In order for plants to become the numerous and large life forms they are on earth today, they required several adaptations from their early beginnings. Describe these adaptations and how they helped plants make their move to land.

4. In the lab, you successfully transformed some bacteria that were *lac*⁻ and therefore couldn't hydrolyze lactose or X-Gal into *lac* + bacteria that could utilize those substrates. Your instructor now challenges you to insert a gene from a jellyfish into the *E. coli* bacteria, which will make them glow in the dark. You have access to the 4500 base pair plasmid you used before which contains the *lac* gene and an ampicillin resistence gene, as well as jellyfish and *E. coli* and all the ancillary materials you need. You are also given the sequence data for the 250 base pair jellyfish gene. You have sixty days to complete your project as well as a small expense account for materials not provided.

    (A) Design an experiment that will enable you to make *E. coli* glow in the dark. Make sure you give a complete protocol and include controls as needed.

    (B) A student got the following results after running a gel to examine the plasmid DNA from different populations of bacteria used in the experiment.
    Lane 1   original plasmid (not inserted)
    Lane 2   original plasmid cut with restriction enzyme (not inserted)
    Lane 3   plasmid from population that was not glowing
    Lane 4   plasmid from population that was glowing
    Lane 5   DNA size standards

Peterson's AP Success: Biology

Answer the following questions based on the above gel and observations that follow:

1. Why were there two different size bands with the uncut plasmid in Lane 1?

2. What is the approximate size for the DNA in Lane 4?

3. Why is the DNA in Lane 4 larger than that in Lane 3?

4. How could some of the bacteria be alive and growing but not glowing?

ANSWERS AND EXPLANATIONS

QUICK-SCORE ANSWERS									
1. C	13. A	25. A	37. B	49. D	61. B	73. B	85. D	97. E	109. B
2. B	14. A	26. C	38. A	50. E	62. C	74. E	86. C	98. C	110. D
3. A	15. C	27. E	39. D	51. E	63. B	75. E	87. D	99. A	111. D
4. E	16. D	28. C	40. D	52. B	64. B	76. C	88. D	100. D	112. B
5. B	17. C	29. C	41. B	53. A	65. B	77. D	89. E	101. C	113. B
6. A	18. E	30. C	42. B	54. C	66. B	78. A	90. B	102. D	114. B
7. D	19. D	31. C	43. C	55. B	67. D	79. B	91. A	103. B	115. E
8. D	20. A	32. D	44. E	56. A	68. A	80. C	92. B	104. A	116. D
9. E	21. B	33. B	45. D	57. D	69. C	70. B	93. C	105. C	117. C
10. B	22. D	34. E	46. A	58. D	70. B	71. A	94. A	106. A	118. D
11. C	23. B	35. B	47. B	59. E	71. A	82. E	83. E	95. E	107. E	119. A
12. D	24. E	36. D	48. C	60. D	72. E	84. B	96. C	108. B	120. E

SECTION I

1. **The correct answer is (C).** Surface tension, in part, is the result of hydrogen bonding and cohesion. Hydrogen bonding is described in choice (C).

2. **The correct answer is (B).** Bond strengths for covalent bonds range from 50–110 kcal/mol (kilocalories per mole), ionic and hydrogen bonds are both around 3–7 kcal/mol, and van der Waals is about 1 kcal/mol.

3. **The correct answer is (A).** The molecule represents glycine, the simplest amino acid. Divided into three parts, the first is the amine group, the second is the carbon backbone with the R group (hydrogen in this case), and the last is the carboxyl terminal.

4. **The correct answer is (E).** See explanation above.

5. **The correct answer is (B).** The sequence of amino acids determines the primary structure, while the regular spacing of hydrogen bonds determines secondary structures, like beta sheets. The interaction of subunits determines quaternary structure.

6. **The correct answer is (A).** The structure of DNA has been referred to as a spiral staircase with the nitrogen bases as the rungs between sugar-phosphate rails or backbones. The nitrogen bases in DNA are adenine, cytosine, guanine, and thymine.

7. **The correct answer is (D).** Any dilution can follow the form $V_1M_1 = V_2M_2$ where M is molarity and V is volume. In this case, solving for V_1, we get 0.04 ml.

$$(2 \text{ ml})(0.2M)/10M = 0.04 \text{ ml}$$

8. **The correct answer is (D).** The proton motive force comes from the electron transport chain in mitochondria and the light reactions in photosynthesis. The segregation of protons is in the inner membrane space in mitochondria and the thylakoid compartment in the chloroplasts.

9. **The correct answer is (E).** The rate of the reaction is dependent on several things in the AP lab, and the choices are either enzyme or substrate concentration. The graph reflects what would happen with increasing quantities of substrate. On the other hand, if one added larger and larger amounts of enzyme, the slope of the line would be constant, just steeper as more is added.

10. **The correct answer is (B).** Prokaryotes lack nuclei and the rest of the organelles mentioned. The differences in our ribosomes is the point of attack for some antibiotics.

11. **The correct answer is (C).** The others are listed in the correct order from A to E for exportation.

12. **The correct answer is (D).** Microtubules serve a support and movement function, participate in mitosis and meiosis, but play no part in muscle contraction. Micro*filaments* are responsible for muscular contraction.

13. **The correct answer is (A).** Because of its lipid nature, only small lipids, or non-polar substances like O_2, can pass through the membrane freely. This is one problem with taking steroid hormones—they pass freely and react within the nucleus.

14. **The correct answer is (A).** Cyclins build up during the cell cycle but do not phosphorylate anything by themselves. They combine with cyclin dependent kinases to form MPF or maturation promoting factor, which is sometimes referred to as M-phase promoting factor.

15. **The correct answer is (C).** Colchicine destroys the microtubules, which comprise the spindle fibers on which chromosomes move during mitosis. With the tubules destroyed, the chromosomes have nowhere to go, and the cell has twice the normal number of chromosomes.

16. **The correct answer is (D).** Mitosis begins with the disintegration of the nuclear membrane, the condensation of the chromosomes, and the synthesis of the spindle apparatus, which define prophase.

17. **The correct answer is (C).** As the fluid mosaic model explains, the embedded proteins in the phospholipid bilayer may move about, floating, as it were, amongst the phospholipids.

18. **The correct answer is (E).** All cell reproduction uses mitosis, which does not reduce the chromosome number. The generation of sex cells requires the reduction of that number (by meiosis).

19. **The correct answer is (D).** Interphase consists of both the G1 and G2 phase of the cell cycle as well as DNA replication. The only time one can see individual chromosomes or pairs of chromosomes is during one of the stages of mitosis.

20. **The correct answer is (A).** The prefix "allos" means "other" or "elsewhere." An inhibitor slows down the rate of the reaction; therefore, an allosteric inhibitor slows it down from elsewhere or by acting elsewhere. Competitive inhibitors are in competition with the substrate for the active site.

21. **The correct answer is (B).** This experiment is a restatement of Peter Mitchell's demonstration that ATP is produced by hydrogen ions that flow down a concentration gradient through ATP synthase. They are originally concentrated in the intermembrane space by the power of the electron transport chain and then flow through ATP synthase, phosphorylating ADP, into the matrix.

22. **The correct answer is (D).** Much of the original energy gained in cellular respiration comes from oxidative phosphorylation when the NADH and $FADH_2$ transport their electrons to the electron transport chain.

23. **The correct answer is (B).** Alcohol fermentation produces alcohol. Ethanol is, of course, a two-carbon alcohol. Fermentation also produces NAD^+, which is necessary for glycolysis.

24. **The correct answer is (E).** The most important product of glycolysis is ATP. Although two ATP are required in glycolysis, four are generated, for a net gain of two. In addition, much important energy is captured by NADH to be used later in the electron transport chain.

25. **The correct answer is (A).** NADH is oxidized back to NAD^+, a necessary component in the oxidation of glucose in glycolysis. Without NAD^+, the cell would have no oxidizing agent and would be unable to gain any ATP from glycolysis.

26. **The correct answer is (C).** The final electron acceptor in cell respiration is oxygen. Electron acceptors are oxidizing agents.

27. **The correct answer is (E).** Phosphofructokinase is an important molecule in the control of glycolysis. There is no need for glycolysis if the cell has ample energy; therefore, molecules like ATP and citrate serve as inhibitors to glycolysis by inhibiting phosphofructokinase. On the other hand, activators like ADP stimulate phosphofructokinase when there is a need for more energy. Kinases usually add phosphates to compounds, and in this case, fructose phosphate is phosphorylated to fructose bisphosphate.

28. **The correct answer is (C).** Water is split, and oxygen is given off in the first stages of the light reaction. The electron acceptor $NADP^+$ becomes the energy rich NADPH, which provides much of the reducing power for the Calvin cycle.

29. **The correct answer is (C).** Rubisco is the most abundant protein on Earth. It serves as the acceptor for carbon dioxide and the fixation of that carbon into sugars.

30. **The correct answer is (C).** There is a one-to-one ratio between oxygen and carbon dioxide in the photosynthesis.

31. **The correct answer is (C).** The person is female because of the two X chromosomes. She has Down syndrome, trisomy 21, because there are 3 21st chromosomes.

32. **The correct answer is (D).** A normal female has 23 pairs of homologous chromosomes. The trisomy at 21 resulted from nondisjunction, where a pair of homologous chromosomes failed to separate in gamete formation. Therefore, considering the three as A, B, and C, A and B are homologous, A and C are homologous, and B and C are homologous.

33. **The correct answer is (B).** Meiosis reduces the normal 2N chromosome number to N chromosome number by cell division. After meiosis I, there are 2 cells, and after meiosis II, there are 4. The number of

chromosomes is rather tricky because the terminology follows the number of centromeres. Meiosis I separates homologous chromosomes, not sister chromatids.

34. **The correct answer is (E).** Changing the number of chromosomes can increase the variation within a population, especially in plants. Crossing over is the physical exchange of genetic material between homologous chromosomes during meiosis I and is the most common means of increasing variation (of those listed).

35. **The correct answer is (B).** The dark ovals represent a pair of chromatids and the light their homologues. Crossing over is the exchange of material between homologous chromosomes.

36. **The correct answer is (D).** The beakers represent the homologous chromosomes. The results show how independent assortment can produce many different combinations. The number of combinations that could arise in the human population is 2^{23}, over 8 million.

37. **The correct answer is (B).** In this dihybrid cross, like others, one may deal with one trait at a time. To have half gray and half red, the parents were Gg X gg. Since three fourths of the offspring were straight, Ss X Ss would be the genotypes for straight/curly.

38. **The correct answer is (A).** Alleles A and B are both dominant to O. Since the child is type O, he had to receive an O allele from each parent. Parents A must have had AO and BO.

39. **The correct answer is (D).** The woman carries one normal and one color-blind gene on her X chromosomes. She will give half of her male offspring the normal and half the color-blind gene, on average. Furthermore, half of the children born to any parents should be boys, on average. The probability of two independent events is the product of their individual probabilities.

40. **The correct answer is (D).** Since the only way to have color is to be *cc*, one fourth of all offspring will have color. Three fourths of those will be green, and one fourth will be yellow.

41. **The correct answer is (B).** Epistasis is the action of one gene on the expression of another. It arises in non-Mendelian ratios, such as above or 9:3:4.

42. **The correct answer is (B).** Each of the other steps was an important clue for them to use. The Meselson-Stahl experiment came later and dealt with how DNA is replicated.

43. **The correct answer is (C).** DNA does not contain sulfur and, therefore, choices (B) and (D) are patently false. Most of the labeled phosphorus was associated with the bacterial cells in their first experiments. When the experiment was lengthened, viral progeny were collected and found to contain the radioactive phosphorus, proving that DNA is handed down from generation to generation.

44. **The correct answer is (E).** The information of DNA is stored in its sequence of base pairs. The RNA strand is polymerized from the DNA strand, and complementary bases are used, A-U, and G-C. Uracil is used rather than thymine in RNA polymerization.

45. **The correct answer is (D).** Restriction enzymes cut DNA at specific sites. The likelihood of finding any base in a DNA sequence is 1 in 4 or 1/4. The probability of finding 6 specific bases is $(1/4)^6$ or 1/4096. Therefore, 3 million divided by 4096 is 732.

46. **The correct answer is (A).** Structure X represents a strand of RNA, which is transcribed in the nucleus.

47. **The correct answer is (B).** The transfer RNA (tRNA) that translates the mRNA strand does so by Watson-Crick base pairing rules. The complement of CAC is GUG, and the amino acid that is transferred by GUG is histidine.

48. **The correct answer is (C).** The biochemical process represented is translation, the making of proteins from the information found in RNA. The central dogma of molecular biology is that DNA is *transcribed* into RNA, which is *translated* into protein. This translation happens at the ribosome, in both prokaryotes and eukaryotes.

49. **The correct answer is (D).** The number of tRNAs is determined by the number of codons and the different amino acids that they ferry. Seemingly, 61 codons need separate tRNAs, but because of wobble, only 45 different tTRAs are required.

50. **The correct answer is (E).** All of the other choices have to do with *eukaryotic* gene expression.

51. **The correct answer is (E).** Proto-oncogenes were first discovered by Varmus and Bishop, who won a Nobel Prize for associating the oncogenes in retroviruses to naturally occurring growth and cell division genes in our cells. The change from proto-oncogene to oncogene is a mutation. These may be point mutations or changes in the chromosome.

52. **The correct answer is (B).** By using the same restriction enzyme, overlapping fragments of DNA will bind with complementary strands that are overlapping from the other type of DNA.

53. **The correct answer is (A).** The heterotroph hypothesis says that the earliest life forms used each other for nutrition, either as predators or decomposers.

54. **The correct answer is (C).** The question revolves around adaptive radiation versus convergent evolution. Adaptive radiation begins with organisms that are closely related but differentiate into different niches and become dissimilar. Convergent radiation begins with organisms that are not closely related but, because of the environment, come to look similar. An example of the latter would be bats and birds.

55. **The correct answer is (B).** Structural and biochemical homologies are used by taxonomists and phylogeneticists to help classify organisms, based on their evolutionary history. The wings of bats and birds do not show relatedness as much as the force of the environment in shaping organisms to fill certain niches.

56. **The correct answer is (A).** Stable gene frequencies would preclude change. Use and disuse and Lamark's theory of acquired characteristics are patently false. Evolution is the change in structure or gene frequencies over time.

57. **The correct answer is (D).** Darwin had read Lyell's book, which gave him the geologic time required for evolution, but he knew nothing of genetics or mutations. The concept of overproduction is one of the cornerstones for his theory for natural selection.

58. **The correct answer is (D).** Punctuated equilibrium theorizes that no changes happen for long periods of time, and then, usually because of environmental catastrophes, short periods of rapid changes take place.

59. **The correct answer is (E).** The Hardy-Weinberg theorem says that gene frequencies will remain constant unless there is mutation, gene flow, natural selection, sexual selection, or genetic drift. The converse realizes that these factors change gene frequencies and, therefore, are the causes of evolution.

60. **The correct answer is (D).** To find the phenotypic frequencies in the F1, multiply the gene frequencies of each type of parent. The gene frequency for red is 0.6 and white is 0.4. Therefore, the frequency of pink is 2rw (since there are 2 ways to make the heterozygote).

	0.6r	0.4r
0.6r	0.36rr	0.24rw
0.4w	0.24rw	0.16ww

61. **The correct answer is (B).** To find the gene frequency, use the Hardy-Weinberg theorem. The genotypic frequency of the homozygous recessives, q^2, is 0.14. To find q, or the frequency of the recessive allele, take the square root of q^2.

62. **The correct answer is (C).** Isolation is the main cause of speciation. While isolation is the most common cause of speciation and geographical the most common isolation type, isolation can also be temporal, behavioral, or mechanical (reproductive).

63. **The correct answer is (B).** This Lamarkian idea of acquired characteristics conflicts with Darwin's theory of natural selection. There has never been any evidence to prove Lamark's theory. While Arnold Schwarzenegger may have a barrel chest developed by lifting weights, his children will not be born that way (although they may have the genetic propensity to develop large muscle mass with work of their own).

64. **The correct answer is (B).** Viruses are not classified as living organisms because they lack cellular structure and can function only in a living cell.

65. **The correct answer is (B).** The 1 mm distance is the same as 1000μm. The small, dark nucleus has a length of one tenth that size.

66. **The correct answer is (B).** While they are both prokaryotes, and therefore lack membrane-enclosed organelles, only Eubacteria have peptidoglycan in their cell walls.

67. **The correct answer is (D).** Parasites obtain their nutrients from living organisms, autotrophs make their own food, and anaerobes live best without oxygen.

68. **The correct answer is (A).** Mitochondria, according to the endosymbiotic theory, developed within ancestral eukaryotes. They, as well as chloroplasts, are not found in prokaryotes.

69. **The correct answer is (C).** In addition to using pseudopodia to capture food, amoebas utilize cytoplasmic streaming for the same function as our circulatory system.

70. **The correct answer is (B).** The contractile vacuole helps this protist deal with the excess water that would accumulate because of its environment.

71. **The correct answer is (A).** Protozoans have been classified according to their method of locomotion.

72. **The correct answer is (E).** Here is one of several mnemonics that people have used to remember the correct order: **King Phillip Came Over For Good Spaghetti.**

73. **The correct answer is (B).** These characteristics define the group Cnidaria (jellyfish, corals, and anemones). The only other choice would have been Ctenophora (comb jellies), which were not listed.

74. **The correct answer is (E).** Coelom is a fluid-filled body cavity that is completely lined by mesoderm tissue. Pseudocoelomates have a body cavity, but the tissues surrounding it were not derived from mesoderm. The only other significant phyla that is pseudocoelomatic is the Rotifera.

75. **The correct answer is (E).** These are the characteristics that define protostomes. Deuterostomes have radial and indeterminate cleavage, mouth from a secondary opening, and coelom formation from the archenteron.

76. **The correct answer is (C).** Conifers belong to the "naked seeded" Coniferophyta. Other major groups of seedless plants are the ferns and whiskferns.

77. **The correct answer in (D).** Chitinous implies arthropods, setae are used by earthworms, chordates have an endo, not exoskeleton, and hydrostatic skeletons are used by organisms like earthworms, who lack either endoskeletons or exoskeletons.

78. **The correct answer is (A).** Nephrons are the function unit in human kidneys. Tracheal tubes and spiracles are respiratory organs in arthropods. Malpighian tubules are excretory organs found in insects.

79. **The correct answer is (B).** Surface area plays a large role in digestion. Maximizing the surface area of food to be digested allows for higher enzyme efficiency, and maximizing the surface area of the digestive tube allows for more area for absorption.

80. **The correct answer is (C).** Homeostasis, the maintenance of the status quo, is an integral theme in biology. Much of the energy organisms use is to maintain their homeostasis. Many "fail-safe" systems, like fainting, are homeostatic mechanisms. Even temperature maintenance in endoderms is homeostatic.

81. **The correct answer is (A).** In the alternation of generations between gametophyte and sporophyte, there has been a move toward dominant sporophyte generation. The most notable plants with dominant gametophyte generations are mosses and liverworts.

82. **The correct answer is (E).** Monocots usually have fibrous root systems, while dicots usually have taproots. The rest of the choices are characteristics that define the monocotyledons, or one seed leaf group.

83. **The correct answer is (E).** Adhesion is the attraction between water and the vessel. Cohesion is the attraction between water molecules that allows them to be pulled up by the driving force of *transpiration*, not root pressure. Companion cells are associated with phloem, not xylem.

84. **The correct answer is (B).** CAM plants only open their stomata at night. While C4 plants use rubisco, it is inside the bundle sheath cell and not close to the atmosphere, where it might use oxygen in photo-

respiration. Mild climates have an abundance of C3 plants, which are not adapted for high heat and low moisture.

85. **The correct answer is (D).** Neurons act on effectors, which are muscles, glands, and other nerves.

86. **The correct answer is (C).** Dendrites serve to receive the incoming messages. Cell bodies are where neurotransmitters are produced and where the nuclei reside. The nodes of Ranvier are nodes on the myelin sheath, which allow the impulse to travel faster.

87. **The correct answer is (D).** This question illustrates the necessity of care in choosing your answer. The most obvious answer was the first one, but one should always peruse the choices.

88. **The correct answer is (D).** Hormones usually travel through the blood. They act on specific tissues called target tissues, which have receptors for the hormones.

89. **The correct answer is (E).** Oxytocin is produced in the hypothalamus and stored for release by the posterior pituitary.

90. **The correct answer is (B).** The right side of the heart is the pulmonary side, which serves the lungs. The left side is the systemic side and serves the rest of the body.

91. **The correct answer is (A).** The letter G represents the atrioventricular valve, whose function is to prevent the backflow of blood into the atrium when the ventricle contracts.

92. **The correct answer is (B).** The letter I is closest to the wall of the right atrium, where the pacemaker is embedded. It is close to where the superior vena cava enters the right atrium.

93. **The correct answer is (C).** Homeostasis is an extremely important concept in biology. See the discussion in the answer to question 80. Myofibrils are units of muscle fibers.

94. **The correct answer is (A).** Salivary amylase begins the breakdown of carbohydrates in the mouth, often making starches taste sweet since they are being hydrolyzed into sugars.

95. **The correct answer is (E).** The small intestine is where most absorption occurs as well as much digestion.

96. **The correct answer is (C).** The stomach's function is the breakdown of proteins, which takes place in an acidic environment with the aid of pepsin.

97. **The correct answer is (E).** This question serves as a tutorial and as a reminder to read all the questions.

98. **The correct answer is (C).** Diagram B represents an alveolus, which is a cluster of air sacs at the end of the smallest bronchioles.

99. **The correct answer is (A).** Number two is the trachea, and the esophagus is behind or dorsal to it.

100. **The correct answer is (D).** The ducts between the kidney, where urine is produced, and the urinary bladder, where it is stored, are the ureters. The urethra is the tube that allows the bladder to empty.

101. **The correct answer is (C).** Ligaments join bone to bone, and tendons join bone to muscle.

102. **The correct answer is (D).** The radius is the forearm bone that rotates as you roll your arm over. Proximal means closest to the midline of the body, and distal means furthest away.

103. **The correct answer is (B).** Class II MHC proteins are involved with the interaction, not class I. Class I MHC proteins are found on all nucleated cells in the body and serve as self markers. They are essential in the cytotoxic T-cell response.

104. **The correct answer is (A).** Antibodies are produced by plasma cells, which are differentiated B cells.

105. **The correct answer is (C).** After ovulation, the follicle becomes the corpus luteum, which secretes progesterone to maintain the endometrium or uterine lining.

106. **The correct answer is (A).** Follicle stimulating hormone (FSH) is produced by the pituitary and acts on the follicles within the ovary. It is inhibited by estrogen and progesterone.

107. **The correct answer is (E).** FSH production is lowest at that time because of the feedback from estrogen and progesterone to the hypothalamus.

108. **The correct answer is (B).** Several major points in a muscle contraction are as follows: first, depolarization by the sodium ions leads to an influx of the calcium stored in the sacroplasmic reticulum; second, calcium ions bind to the troponin, which is attached to the tropomyosin, causing it to twist and reveal the active binding sites for the myosin heads to attach to the actin; third, ATP is required for the relaxation of that binding, not the initiation.

109. **The correct answer is (B).** The ability to become quickly depolarized and repolarized depends on the influx of sodium from outside the membrane, followed quickly by the efflux of potassium to restore the positive charges outside the plasma membrane lost when the sodium rushed inside. During the resting state, the sodium activation gate is closed and the inactivation gate is open. The activation gate of the potassium channel is also closed at this point. When an increase in voltage triggers the sodium channels, the activation gates open, resulting in the depolarizing phase. After a short delay, the sodium inactivation gate closes and the potassium activation gate opens, resulting in repolarization. The undershoot occurs because the potassium channels remain open, while both gates of the sodium channels are closed.

110. **The correct answer is (D).** The amount of biomass is always greater lower down on the food chain or pyramid. This is the result of the loss of energy during transfer from one trophic level to the next.

111. **The correct answer is (D).** Plants obtain about 1 percent of the energy available to them. Consumers on the average obtain 10 percent of the energy available to them.

112. **The correct answer is (B).** Succession is the semi-orderly progression of species replacement.

113. **The correct answer is (B).** Symbiotic relationships are viewed from two perspectives. If both parties benefit, it is mutualism. If one party benefits to the detriment of the other, it is parasitism. If one party benefits and the other is not effected, the relationship is termed commensalism. When both parties are negatively affected, the relationship is usually competition.

114. **The correct answer is (B).** Increasing altitude results in colder temperatures, as does increasing latitude.

115. **The correct answer is (E).** Herbivores depend on green plants for their sustenance. Those that can capture the most with the least amount of energy are usually the most successful. Large size is not a prerequisite for a good competitor.

116. **The correct answer is (D).** Saprophytes obtain their nutrients from dead organisms, while parasites obtain theirs from living organisms.

117. **The correct answer is (C).** The oceans have less temperature change and the most constant abiotic conditions.

118. **The correct answer is (D).** Deciduous forests are named for the trees that shed their leaves once a year due to temperature changes. There is a tropical deciduous forest in Asia in which trees shed because of lack of rainfall.

119. **The correct answer is (A).** Tropical forests are defined by their proximity to the equator, where the average temperature and length of daylight change little during the year, causing an abundance of broad-leaved plants that are competing for the sunlight.

120. **The correct answer is (E).** Olfactory imprinting is thought to be the mechanism for their homing behavior. Habituation would require many trips, operant conditioning is like house-training a dog, and reasoned behavior asks a bit much of a fish.

Section II

1a. Cell cycle and reproduction
The question should be divided into 3 parts.

(A) Proteins that control cell division

- Histones: positively charged proteins to which DNA is bound
- Growth factors and hormones: signal a cell to divide
- Proto-oncogene products: act as both the accelerator and brakes for cell division
- Cyclin-dependent kinases (cdks): concentration remains constant throughout the cycle, but together with cyclin sends the message for the cell to divide.
- Cyclins: as cyclin concentration increases, the cell is moved toward the point of no return for division.
- Regulatory proteins: activated by the combination of cdks and cyclins, called maturation promoting factor (MPFs), degrade cyclins

(B) Proteins used in mitosis

- Centrioles: pairs of 9 triplet microtubules in a ring that have microtubules between them
- Microtubules: the (polar) spindle fibers of mitosis, attach to kinetochores, and used for cell elongation during anaphase
- Centrosome: the micotubule organizing center
- Kinetochores: used to move chromosomes to the midline at metaphase

(C) Proteins used in DNA replication

- DNA polymerases: elongation and replacement of RNA primers

- Helicases: divide the double helix for replication
- Primases: begin replication with RNA to prepare for polymerase
- Single strand binding proteins: hold single strand apart for polymerization
- Nucleases: excise DNA mismatches and other mistakes
- Ligases: link sections of DNA

1b. Cell movement

- motor molecules: protein complexes that interact with one of three different-sized proteins, which include
- microtubules, the thickest filaments, made of tubulin
- microfilaments (actin polymers), the thinnest filaments, bound to plasma-membrane proteins and used for contraction
- integrins: proteins that span the plasma membrane
- collagen: abundant protein attached to integrins, extracellular
- fibronectins: proteins that are bound between collagen and integrins
- myosin: larger protein used for cytoplasmic streaming and ameboid movement (with actin)
- flagella: 9 doublets of microtubules in a ring around 2 single microtubules
- cilia: same as flagella
- dynein: large protein associated with the microtubules in flagella and cilia that helps them move

1c. Signal transduction and communication

- First messenger: peptide hormone that acts on specific (target) tissue
- receptor: protein embedded in plasma membrane that receives hormone
- relay protein: embedded in plasma membrane, stimulate or inhibit; G proteins, for example
- effector protein: acted on by relay, causes change in second messenger; adenyl cyclase, for example
- steroid receptors: found in the nucleus, receive steroid hormones, and bind with DNA or cause regulatory proteins to bind with DNA
- various protein kinases and phosphorylases: acted on by second messengers or third messengers, activate/deactivate reactions and amplify the first message considerably
- calmodulin: combines with calcium to trigger cell responses
- structural proteins, like opsins, initiate many sensory responses
- ion channels: responsible for neuronal function

2a. Types of DNA within cells

prokaryotes:

- DNA without a nucleus; bound to plasma membrane; not associated with histones, circular
- Plasmids: circular pieces of DNA that can be exchanged by prokaryotes

eukaryotes:

- DNA within nucleus; bound to histones, arranged in chromosomes
- Mitochondrial DNA: prokaryotic-like DNA; reproduces on its own schedule; maternally inherited
- Chloroplast or Plastid DNA: same as mitochondrial

2b. Differences in DNA and types of changes

Mutations: major cause and source of most variation

point mutations (change in single nucleotide)

silent: produces no change in protein

missense: produces change in amino acid coded for

nonsense: produces a stop in coding sequence

duplication and deletion: most damaging when the codon reading frame is changed (frameshift mutation)

alterations of chromosome structure

inversions: reverse the order

translocations: moving DNA from one chromosome to another

duplication: chromosomal fragment relocates to homologous chromosome

deletion: chromosomal fragment is lost

Variable number of tandem repeats: 10–15 percent of our DNA is composed of short sequences repeated many times

Transposons: pieces of DNA that move about

Nondisjunction: during meiosis, chromosomes don't separate properly, and an extra chromosome is distributed to one gamete, while another gamete is short a chromosome

Polyploidy: whole sets of chromosome increase chromosome number, common in plants

Mutagenic changes: caused by environmental factors, chemicals, and radiation

2c. Scientists studying changes in DNA

Gel electrophoresis: an electric current causes the negatively charged DNA to move toward a positive pole. The separation of DNA in agarose or polyacriamide gels is based on size (as smaller molecules get through the gel faster).

Restriction enzymes: nucleases that cut DNA usually at palindromic sites; by subjecting different DNAs to the same restriction enzyme and then using gel electrophoresis, scientists can find if there are differences or divide a genome into smaller sizes.

RFLPs: Restriction Fragment Length Polymorphisms; DNA is digested with restriction enzymes (1 or more). Can be used as DNA fingerprint with VNTR and Southern blots

Southern blot: DNA is digested with restriction enzymes, separated by electrophoresis, blotted and denatured, and probed with a radioactive single strand of DNA; probe will attach to complementary regions of the digested DNA; used for gene location, DNA fingerprinting, and gene mapping (many RFLP sites can serve as markers for disease-causing genes).

Polymerase Chain Reaction (PCR): Through different temperature cycles, specific primers, and a heat resistant DNA polymerase, small amounts of DNA can be amplified millionfold and more; used for VNTR and DNA fingerprinting, sequencing, and restriction analyses.

Sequencing: determining the actual base sequence of DNA, usually with PCR and flourescently labeled dideoxyribonucleotides.

In vitro mutagenesis: introduction of specific changes into the sequence of a cloned gene, sometimes referred to as "knock-out" genes

cDNA libraries: complementary DNA is made from mRNA; lacks introns

Gene expression in *Xenopus* eggs

Cloning genes in bacteria and yeast

3. **Background and evolution: early plants most likely descended from a green algae life form, based on storing carbohydrates as starch, pigments, and life cycle similarities. Therefore, they had no roots, true stems, or means to conserve water. They also needed a different means of reproduction if they were going to leave the water.**

Water relations:

- Obtain through roots
- fine root hairs increase surface area and therefore absorption
- roots associated with fungi (mycorrhizae) increase water absorption
- Conservation
- cuticle on leaf surface reduces water loss
- guard cells control stomata, the openings in the leaf
- C4 photosynthesis allows stomata to be closed longer (and therefore less water loss with transpiration) in dry conditions because of increased carbon dioxide assimilation efficiency
- CAM (Crassulacean Acid Metabolism) photosynthesis provides plants with a mechanism for opening their stomata only at night (keeping transpiration to a minimum)
- Conduction
- xylem is the vascular tissue for plants
- phloem transports dissolved sugars, amino acids
- water is transported over long distances because of several forces
- cohesion: the attraction of water molecules to each other so that they are pulled along
- adhesion: the attraction of water moleculed to the xylem wall
- transpiration: the force that creates the negative force or tension that pulls water molecules up the xylem

Reproduction:

- gamete transfer
- sperm swim to ovule or embryo sac in few plants (ferns, for instance)
- sperm may travel in the wind
- sperm may use animals for pollination
- seed production

Peterson's AP Success: Biology

- seeds contain an embryo and a supply of nutrients, protection the embryo from dessication
- fruit production
- ovary develops into a fruit, which both protects and nourishes the embryo
- fruit is often adapted for seed dispersal (coconuts, cockleburrs, and dandelions)

4. There is more than one way to succeed at this experiment.

A. Method 1

- Make *E. coli* cells competent by placing them in cold calcium chloride (or some other method). Keep on ice for up to 48 hours.
- Meanwhile, cut plasmid and jellyfish DNA with the same restriction enzyme.
- Put together and add ligase.
- Later, add to competent cells, heat shock, and ice them.
- Induce growth and plasmid replication by adding nutrient broth.
- Plate on both regular agar and ampicillin agar (should have growth on regular, and those growing on ampicillin have been transformed).
- Further control: plate non-transformed bacteria on ampicillin containing agar (if they grow, then there was not necessarily a transformation in this experiment).

B. Method 2

- Using the gene sequence, design primers and order them.
- Isolate genomic jellyfish DNA.
- Use PCR to amplify gene (with new primers).
- Use appropriate restriction enzyme to cut plasmid and gene.
- Add together and ligase.
- Later, add to competent cells, heat shock, and ice them.
- Induce growth and plasmid replication by adding nutrient broth.
- Plate on both regular agar and ampicillin agar.
- Use other controls as above.

1. The two different size bands reflect a supercoiled structure that runs faster through the gel than linearized molecules and a nicked structure that runs slower.

2. The size of the band in lane 4 is approximately 5,500 base pairs long.

3. The larger size reflects the successful insertion of the gene into the plasmid. It is the combination of the original plasmid (4500 bp) plus the insertion (1000 bp).

4. The bacteria that were alive that were not glowing were transformed by the plasmid that provided them with the ampicillin gene but did not have the insert.

PRACTICE TEST 2

AP BIOLOGY

SECTION I

Time–1 Hour and 30 Minutes
Number of Questions–120
Percent of Total Grade–60

Directions: Each of the questions or incomplete statements below is followed by five suggested answers. Select the one that is best in each case.

1. In the dehydration synthesis reaction to form a triglyceride, how many molecules of water are released?

   (A) one

   (B) two

   (C) three

   (D) four

   (E) The number varies, depending upon the triglyceride formed.

2. Which of the following pairs of structures can be seen in the low-power field of a compound light microscope?

   (A) mitochondria and Golgi body

   (B) cell wall and chloroplasts

   (C) ribosomes and endoplasmic reticulum

   (D) lysosomes and genes

   (E) mitochondria and ribosomes

3. Tracheids differ from vessel elements in that

   (A) water moves between vessel elements by diffusion

   (B) tracheids contain only 25 percent of their original living protoplasm when mature

   (C) water moves between tracheids through pits, not pores

   (D) tracheids have a dual purpose in gymnosperms, carrying both water and food

   (E) tracheid pores occur in groups of two's

4. Meissner's corpuscles are present

   (A) in the epidermis of all mammals

   (B) on body surfaces that do not contain hair

   (C) deeper in the dermis than the Pacinian corpuscles

   (D) wrapped around the hair follicle

   (E) in greatest concentrations when paired with free nerve endings

5. If the ribosomes in a cell ceased to function, which of the following would occur in a cell?

(A) Antibody production would increase.

(B) Mitosis would proceed uncontrollably.

(C) Rate of glucose transport in the cytoplasm would increase.

(D) The sodium-potassium pump would spontaneously start, creating an ionic imbalance between the intercellular and extracellular fluids.

(E) Enzyme synthesis would cease.

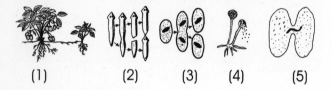

(1)   (2)   (3)   (4)   (5)

6. Which diagram above shows the type of asexual reproduction known as regeneration?

(A) 1

(B) 2

(C) 3

(D) 4

(E) 5

7. All of the following are parts of the axial skeleton **EXCEPT** the

(A) skull

(B) vertebral column

(C) scapula

(D) sternum

(E) rib cage

8. Rhodopsin is a

(A) red pigment found in the rhodophyta

(B) hemoglobin substitute found in the blood of invertebrates

(C) proteolytic intestinal enzyme

(D) photopigment in rods of the retina

(E) hormone that stimulates the development of pigment in hair and fur

9. Glycolysis occurs

(A) in the ribosomes

(B) in the cytoplasm

(C) in the intercellular matrix of the mitochondria

(D) on the cristae of the mitochondria

(E) on the inner surface of the plasma membrane

10. The ampullae of Lorenzini would be found in the

(A) walls of the abdomen of grasshoppers

(B) body cavity of segmented worms

(C) heads of sharks

(D) inner ear of mammals

(E) abdomen of skunks

11. Plants grow by means of their apical meristems, which in turn give rise to three types of primary meristems. These are

(A) protoderm, lateral meristem, and ground meristem

(B) procambium, cork cambium, and ground meristem

(C) procambium, lateral meristem, and protoderm

(D) protoderm, procambium, and ground meristem

(E) lateral meristem, cork cambium, and ground meristem

12. Which of the following nitrogen bases is present only in RNA?

   (A) Guanine

   (B) Cytosine

   (C) Adenine

   (D) Uracil

   (E) Thymine

13. Which organism is correctly paired with its main adaptation for gas exchange?

   (A) amoeba-nucleus

   (B) earthworm-nephridia

   (C) grasshopper-Malpighian tubules

   (D) human-pancreas

   (E) frog-skin

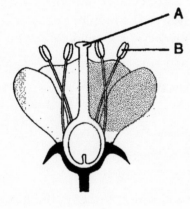

14. The transfer of reproductive structures from B to A is known as

   (A) self-pollination

   (B) oogenesis

   (C) differentiation

   (D) ovulation

   (E) spermatogenesis

15. The patellar reflex involves a sensory and motor neuron in the spinal cord. It is termed a(n)

   (A) autonomic reflex

   (B) parasympathetic reflex

   (C) unconscious reflex

   (D) ganglia reflex

   (E) monosynaptic reflex

16. All of the following are hormones produced by the anterior lobe of the pituitary **EXCEPT**

   (A) antidiuretic hormone (ADH)

   (B) growth hormone (GH)

   (C) thyroid-stimulating hormone (TSH)

   (D) luteinizing hormone (LH)

   (E) follicle-stimulating hormone (FSH)

17. A tumor-suppressor gene can prevent replication of damaged DNA by

   (A) degrading the damaged DNA

   (B) blocking the cell cycle at the G1 checkpoint

   (C) activating a newly transcribed DNA molecule

   (D) initiating lysis of the cell containing the damaged DNA

   (E) methylating the defective nucleotides

18. Ethylene, a simple gaseous hydrocarbon, is a naturally occurring plant hormone. Its production in a plant would be increased by all of the following factors **EXCEPT**

   (A) insecticide spraying

   (B) exposure to ozone

   (C) temperature extremes

   (D) attack by pathogens

   (E) attack by herbivores

19. One type of cell communicating mechanism allows materials to pass through cells and not between them. This is due to the nature of the cells being in direct contact with each other. An example would be found in the epithelial cells of the gut and is known as a(n)

    (A) tight junction
    (B) gap junction
    (C) plasmodesmata
    (D) adherens junction
    (E) desmosome

20. Which of the following is NOT a reason Mendel chose to work with pea plants for his studies of inheritance?

    (A) Pea plants are easy to grow.
    (B) Mendel knew from other work that peas could produce hybrids by breeding different varieties.
    (C) Pea plants do not exhibit sex-linked characteristics when crossbred with other varieties.
    (D) Sexual organs are enclosed within the flower.
    (E) There were a large number of true-breeding varieties available.

21. The acoelomates are the most primitive bilaterally symmetrical animals. Only one type of internal system is common to all acoelomates. This is a _____ system.

    (A) respiratory
    (B) digestive
    (C) nervous
    (D) reproductive
    (E) skeletal

22. All of the following processes transport mechanisms into a cell EXCEPT

    (A) phagocytosis
    (B) contractile vacuoles
    (C) porin channels
    (D) receptor-mediated endocytosis
    (E) pinocytosis

23. Uropod, carapace, and swimmerets are all structures found in

    (A) annelids
    (B) bony fish
    (C) cartilaginous fish
    (D) jawless fish
    (E) crustaceans

24. The series of cellular divisions by which a one cell-layered structure becomes a three-layered embryo is known as

    (A) gastrulation
    (B) cleavage
    (C) blastulation
    (D) meiosis
    (E) nondisjunction

25. During a chemical reaction, the loss of an electron from an atom or molecule is termed

    (A) reduction
    (B) cellular respiration
    (C) kinetics
    (D) oxidation
    (E) metabolism

26. In a closed circulatory system with a four-chambered heart, the point of highest pressure in the system is the

    (A) capillary bed between arterioles and venules
    (B) right atrium
    (C) right ventricle
    (D) left atrium
    (E) left ventricle

27. Haversian canals and a matrix of crystals of calcium phosphate would be found in

    (A) teeth
    (B) nails
    (C) hooves
    (D) bone
    (E) cartilage

28. What happens to chylomicrons found in the blood stream?

    (A) They are gradually broken down to form fatty acids.
    (B) They are actively transported into muscle cells, where they are broken down, and the resulting fatty acids are used for energy.
    (C) They diffuse into fat cells and are stored there for future use.
    (D) They are actively transported into liver cells, where the proteins are lysed and the remaining fatty acids are released back into the blood stream.
    (E) They are actively transported into adipose cells, where the proteins are lysed and the remaining fatty acids are stored.

29. Baroreceptors, located in the walls of the carotid artery and the aorta, respond to a decrease in systemic arterial blood pressure by sending impulses

    (A) to the medulla, which in turn stimulates the sympathetic nerves and increases the rate and force of contractions by the heart ventricles
    (B) to the medulla, which in turn stimulates the parasympathetic nerves and increases the rate and force of contractions by the heart atria
    (C) directly to the left atrium
    (D) directly to the left ventricle
    (E) to the diaphragm to stimulate increased respiration

30. The response by which tendrils curl around and cling to stems and other objects is an example of

    (A) photoperiodism
    (B) cytokinins
    (C) gibberellins
    (D) gravitropism
    (E) thigmotropism

31. Smooth muscle cells lack a sarcoplasmic reticulum

    (A) so during a contraction, Ca^{++} ions enter from the extracellular fluid
    (B) and therefore use Na^{+} ions to stimulate the binding to muscle proteins
    (C) so the Ca^{++} ions travel along the sarcomeres instead
    (D) thereby allowing faster contractions without the aid of ion movement
    (E) and this gives them an appearance of being smooth

32. Control of breathing and heartbeat are located in the

    (A) occipital lobe of cerebrum
    (B) cerebellum
    (C) medulla
    (D) frontal lobe of cerebrum
    (E) pituitary gland

33. Monovalent ions such as Na+ and Cl- are recovered by a vertebrate in which part of the nephron?

    (A) Bowman's capsule
    (B) proximal tubule
    (C) loop of Henle
    (D) distal tubule
    (E) collecting duct

34. Scientists have been able to produce mutations in plants by irradiating their seeds with gamma rays. One of the plants that was produced could not produce flowers. Because of the lack of flowers, this plant would be unable to

    (A) photosynthesize
    (B) transport water
    (C) reproduce sexually
    (D) grow more than a few inches tall
    (E) respond to gravitropic stimuli

35. A major advantage of a closed circulatory system is that

    (A) the flow of fluid in the vessels is regulated by muscle-controlled changes in the diameter of the blood vessels
    (B) oxygen is carried to all parts of the body
    (C) hemoglobin is never in direct contact with body cells
    (D) carbon dioxide can be used as a buffer in the blood
    (E) there is less chance of bleeding to death during a major injury to the body

36. Sodium and potassium pass through the neural membrane by

    (A) movement through channels
    (B) endocytosis
    (C) diffusion
    (D) passive transport
    (E) phagocytosis

37. The following steps are all involved in the immediate response of the body to an invading virus.
    1-secretion of interferons
    2-attack of infected cells by natural killer cells
    3-macrophages engulf infected cells
    4-macrophages enzymatically degrade the protein coat of the virus
    5-macrophages display coat fragments on their surface
    6-T cells recognize viral surface antigens

    The correct order of events is

    (A) 1, 2, 3, 4, 5, 6
    (B) 1, 6, 4, 3, 5, 2
    (C) 1, 6, 4, 5, 2, 3
    (D) 6, 1, 2, 3, 4, 5
    (E) 6, 2, 4, 3, 1, 5

38. Which term is out of place with the other four?

(A) Islets of Langerhans

(B) acinar cells

(C) parietal cells

(D) chief cells

(E) salivary glands

39. The testes are formed in the

(A) inguinal canal but normally drop into the scrotum before birth

(B) inguinal canal but normally drop into the scrotum within three months after birth

(C) abdominal cavity, but normally drop into the scrotum through the inguinal canal before birth

(D) abdominal cavity but normally drop into the scrotum through the inguinal canal within three months after birth

(E) scrotum, which forms outside the abdominal cavity in the first trimester of pregnancy

40. Artificial selection is illustrated by

(A) random mating that occurs in a population

(B) the appearance of a new species on an isolated island

(C) a cattleman producing a larger cow by breeding specific cows to a prize bull

(D) a gardener choosing certain flowers to cut to display at a flower show

(E) wind assisting the pollination of grass in a field

41. In noncompetitive inhibition, the allosteric inhibitor

(A) attaches to the active site, preventing the substrate from attaching there

(B) attaches to the substrate, preventing it from attaching to the active site

(C) changes the pH of the environment, thus preventing enzyme-substrate complex formation

(D) causes the substrates to polymerize, preventing individual enzyme-substrate attachment

(E) attaches to the enzyme at a site away from the active site, altering the shape of the enzyme

42. The very acid pH in the stomach, about 2.0, causes all of the following **EXCEPT**

(A) activation of pepsinogen

(B) emulsification of fats

(C) denaturation of food proteins

(D) high death rate due to bacteria in the stomach

(E) maximum activity of pepsin

43. Place the steps in the evolution of metabolism in their proper sequence.

1–development of organized sequences of catalyzed degradation reactions

2–photosynthesis using H_2S

3–ability to harness chemical bond energy as in the bonds of ATP

4–substitution of H_2O for H_2S in photosynthesis

5–development of a mechanism for fixing nitrogen

6–appearance of aerobic respiration

(A) 3, 2, 4, 1, 6, 5

(B) 3, 1, 2, 5, 4, 6

(C) 5, 3, 4, 3, 6, 1

(D) 1, 2, 5, 6, 4, 3

(E) 1, 3, 5, 2, 4, 6

44. Cuboidal epithelium would be found

(A) in the walls of the heart

(B) in the knees and other joints as well as in the nose and tracheal rings

(C) on the surface of the skin

(D) lining glands and kidney tubules, and covering the ovaries

(E) in tendons and the sheath around voluntary muscles

45. The pedigree chart below shows the pattern of inheritance for a sex-linked trait.

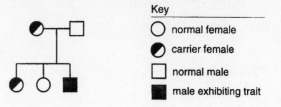

If this couple has another son, what is the probability that he will exhibit this sex-linked trait?

(A) 0 percent

(B) 25 percent

(C) 50 percent

(D) 75 percent

(E) 100 percent

46. Of the following, all are considered appendages **EXCEPT**

(A) pedipalps

(B) antennae

(C) telson

(D) swimmerets

(E) cheliped

Peterson's AP Success: Biology

47. What would be the mRNA triplet complementary to the tRNA molecule shown in the diagram below?

T-RNA Molecule

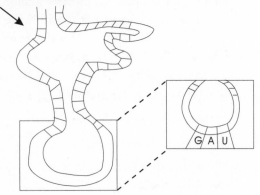

(A) CTA

(B) GAU

(C) CUA

(D) UAG

(E) CUT

48. Of the following, the structures which are analogous to the bird wing are

(A) a dog forelimb

(B) a butterfly wings

(C) a human arm

(D) a whale fin

(E) a frog forelimb

49. Characteristics of a species that make its members better able to live and reproduce in their environment are known as

(A) acquired characteristics

(B) homologous structures

(C) abiotic factors

(D) biotic factors

(E) favorable adaptations

50. The polarity of the water molecule causes it to be attracted to other polar molecules. This property is referred to as

(A) adhesion

(B) cohesion

(C) surface tension

(D) specific heat

(E) hydrogen bonding

51. The presence of chlorophyll *a* and chlorophyll *b* would place an unknown protist into which of the following phyla?

(A) Rhodophyta

(B) Chlorophyta

(C) Ciliophora

(D) Phaeophyta

(E) Oomycota

52. The primary reason that amphibians never succeeded in becoming fully terrestrial is

(A) their lungs were not large

(B) ancestral amphibians did not use their skin as a respiratory organ

(C) the presence of terrestrial predators, causing them to remain relatively safe in water environments

(D) they must lay their eggs in water to avoid them drying out

(E) their three-chambered heart was not an efficient terrestrial circulatory pump

53. An insufficient amount of hemoglobin is most closely associated with the disorder known as

(A) angina

(B) anemia

(C) coronary thrombosis

(D) arteriosclerosis

(E) hypotension

54. A large population of houseflies was sprayed with a newly developed, fast-acting insecticide. The appearance of some houseflies that are resistant to this insecticide supports the concept that

(A) species traits tend to remain constant

(B) biocides cause mutations

(C) fast-acting insecticides are not 100 percent effective in killing houseflies

(D) variation exists within a species

(E) the environment does not change

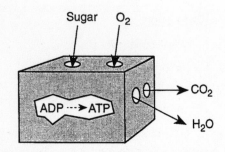

55. In which organelle would the events pictured above most likely occur?

(A) mitochondrion

(B) chloroplast

(C) lysosome

(D) centriole

(E) ribosome

56. Glyphosate is the active ingredient in herbicides and kills plants by inhibiting an enzyme required by plants to produce aromatic amino acids. The effect of glyphosate on humans is

(A) a slow paralysis of the vocal cords

(B) tetany of the involuntary muscles

(C) sterility

(D) depression

(E) no effect at all

57. A physiologist removed the pancreas from several dogs in an experiment to investigate the function of the pancreas. He placed five normal dogs in one kennel and five dogs lacking a pancreas in another kennel. The physiologist observed that ants were attracted in large numbers to the kennel of the dogs lacking a pancreas. Because they lacked a pancreas, what substance did these dogs have that attracted the ants?

(A) enzymes in their saliva

(B) sugar in their urine

(C) mineral salts in their feces

(D) oil on their fur

(E) keratin in their claws

58. The large demands for oxygen from the flight muscles of a bird have resulted in

(A) a more efficient energy transport chain

(B) the largest lung-to-body-mass ratio in the vertebrates

(C) storage of extra oxygen in the hollow bones for use in flight

(D) an efficient three-chambered heart

(E) a respiratory system comprised of anterior and posterior air sacs in addition to the lungs

59. Leukocytes are formed in the bone marrow

(A) and develop in the bone marrow as T and B cells

(B) from the same stem cells as red blood cells

(C) but some migrate to the thyroid and develop as T cells

(D) and lack a nucleus until they ingest their first pathogen, whereupon they are stimulated to develop a nuclear envelope around their chromosomes

(E) but those traveling to the lymph nodes are transformed into lymphocytes

60. Which structures control the cyclic nature of the menstrual cycle?

(A) oviduct and uterus

(B) pituitary and testes

(C) ovaries and umbilical cord

(D) pituitary and ovaries

(E) oviduct and thyroid

61. During tissue development in vertebrates, the primary cell layers begin their transformation and the notochord develops from

(A) mesoderm along the dorsal midline

(B) endoderm along the dorsal midline

(C) somites lateral to the site of the notochord

(D) ectoderm above the neural groove

(E) the primitive streak closing

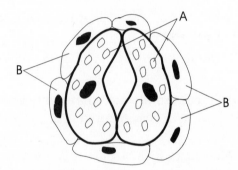

62. In the above diagram, the opening between the two 'A' cells will close due to

(A) movement of oxygen from the cells surrounding the 'A' cells into them.

(B) movement of water out of the 'A' cells

(C) accumulation of glucose in the 'A' cells during the dark phase of photosynthesis

(D) accumulation of carbon dioxide in the 'B' cells

(E) a decrease in the environmental temperature

63. Haversian canals and lamella would be found in

(A) mollusc shells

(B) vertebrate bones

(C) dicot stems

(D) monocot leaves

(E) the inner ear of vertebrates

64. Interaction between genes where the product of one gene can modify the phenotypic expression of another gene is

(A) epistasis

(B) co-dominance

(C) incomplete dominance

(D) pleiotropy

(E) dihybridization

65. Hormones produced by the testes control the expression of traits for

(A) hair color and eye color

(B) beard development and number of fingers

(C) hair color and voice quality

(D) number of fingers and eye color

(E) voice quality and beard development

66. A hydra is considered more advanced than a sponge because the

(A) hydra has extracellular digestion of food.

(B) hydra has tentacles that contain stinging nematocysts.

(C) hydra has a mouth that points to the base of its body.

(D) harpoon of the nematocyst uses osmotic pressure.

(E) hydra's body lacks the pores that are present in a sponge.

67. Which of the following occurs during DNA replication?

(A) DNA polymerase adds nucleotides to the 5' end only.

(B) The leading strand elongates away from the replication fork.

(C) Okazaki fragments are made in a 3' to 5' direction.

(D) Both new strands of the DNA are replicated in the same direction.

(E) An RNA primer must start replication since DNA polymerase cannot link the first nucleotides in a newly synthesized strand.

68. In human males, sperm cells are suspended in a fluid medium. The main advantage gained from this adaptation is that the fluid

(A) removes polar bodies from the surface of the sperm

(B) activates the egg nucleus so that it begins to divide

(C) acts as a transport medium for sperm

(D) provides currents that propel the egg down the oviduct

(E) prevents deformed sperm or those containing mutations from fertilizing the egg

69. Which process is illustrated by the arrows in the diagram below?

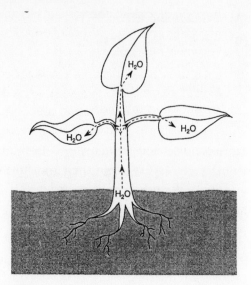

(A) enzymatic hydrolysis

(B) cytoplasmic streaming

(C) intracellular circulation

(D) intercellular transport

(E) active transport

70. Some of the events that occur during the synthesis of a specific protein are listed below.
1–Messenger RNA attaches to a ribosome.
2–DNA serves as a template for RNA production.
3–Transfer RNA bonds to a specific codon.
4–Amino acids are bonded together.
5–RNA moves from the nucleus to the cytoplasm.

The correct order of these events is

(A) $2 \rightarrow 5 \rightarrow 1 \rightarrow 3 \rightarrow 4$

(B) $4 \rightarrow 1 \rightarrow 5 \rightarrow 3 \rightarrow 2$

(C) $2 \rightarrow 3 \rightarrow 5 \rightarrow 4 \rightarrow 1$

(D) $5 \rightarrow 1 \rightarrow 3 \rightarrow 4 \rightarrow 2$

(E) $3 \rightarrow 2 \rightarrow 1 \rightarrow 5 \rightarrow 4$

Peterson's AP Success: Biology

71. Water functions as a powerful solvent of polar molecules because

(A) water molecules tend to form the maximum number of polar bonds between molecules

(B) each molecule of the solute that breaks loose from the existing solid crystal is surrounded by a hydration shell, preventing the solute molecule from reassociating with other solute molecules

(C) the excess hydrogen ions act like an acid to dissolve the solute

(D) the hydrophobic nature of water causes solute molecules to be attracted to water

(E) polar molecules have a strong tendency of repelling each other, allowing the charged portions of the water molecule to bond with these polar molecules

72. The diagram below illustrates what happens to the fur coloration of a Himalyan hare after exposure to a low temperature.

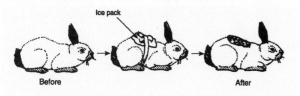

Before   After

This change in fur coloration is most likely due to

(A) the arrangement of genes on homologous chromosomes

(B) a sex-linked environmental allele

(C) mutations resulting from a change in the environment

(D) the effect of heredity on gene expression

(E) environmental influences on gene action

73. Barr bodies are

(A) inactivated X chromosomes in females

(B) one of the genetic mutations in *Drosophila*

(C) one of the physical manifestations of trisomy-21

(D) the physical attachment of actin in a sarcomere

(E) the dark lines that appear across chromosomes

74. The first group of organisms to evolve a body cavity are the

(A) cnidarians

(B) nematodes

(C) platyhelminthes

(D) annelida

(E) molluscs

75. The enzymes below are all secreted into the alimentary canal.
1–amylase
2–sucrase
3–pepsin
4–trypsin

The correct order of their secretion is:

(A) 1, 2, 3, 4

(B) 1, 3, 2, 4

(C) 4, 1, 3, 2

(D) 3, 4, 1, 2

(E) 3, 1, 4, 2

Questions 76 through 78 refer to the diagram below.

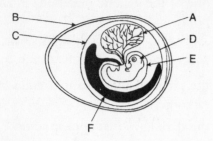

76. Which structure provides nourishment for the developing organism?

(A) A

(B) B

(C) C

(D) D

(E) E

77. Which structure provides a watery environment and protects the embryo from physical shock?

(A) A

(B) B

(C) C

(D) D

(E) E

78. Letter C indicates the

(A) allantois

(B) chorion

(C) amnion

(D) yolk sac

(E) placental membrane

79. The diagram below shows the bones in an appendage of a cat and in an appendage of a bird.

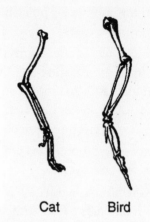

Cat   Bird

The similarities between these appendages suggest that cats and birds

(A) possess the same number of chromosomes

(B) evolved from a common ancestor

(C) evolved gradually over a long period of time

(D) evolved during a short burst of rapid evolutionary change

(E) occupy the same habitat

80. When a species shows long periods of stability interrupted by geologically brief periods of significant change, this is referred to as

(A) use and disuse

(B) evolutionary gradualism

(C) reproductive isolation

(D) homologous structures

(E) punctuated equilibrium

81. What is the complementary messenger-RNA sequence for the DNA sequence shown below?

T-A-C-G-G-C-A-A - T

(A) T-A-C-G-G-C-A-A - T

(B) A-T-G-C-G-C-T-T - A

(C) U-T-G-C-G-C-T-T - U

(D) U-T-C-G-G-C-T-T - A

(E) A-U-G-C-C-G-U-U - A

82. The G_o phase

(A) includes all of the mitotic stages

(B) is a resting phase that occurs prior to the G_1 phase

(C) is a resting phase that occurs during the G_1 phase, prior to possibly resuming division

(D) occurs only in rapidly dividing cells

(E) occurs after G_2 but just prior to mitosis, allowing the cell to prepare for the actual nuclear division

83. The cytoskeleton is made of

(A) actin filaments

(B) actin and myosin filaments

(C) actin filaments and microtubules

(D) actin and myosin filaments and microtubules

(E) actin and intermediate filaments and microtubules

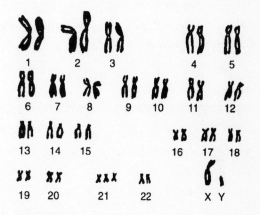

84. The genetic disorder in the diagram above most likely resulted from

(A) crossing-over

(B) nondisjunction

(C) polyploidy

(D) deletion

(E) segregation

85. Which of the following is a characteristic of deuterostomes?

(A) The mouth forms from the blastopore.

(B) The coelom arises from a mesodermal split, not from gut outpouching.

(C) Cleavage is spiral, not radial.

(D) The developmental fate of each cell is not fixed from the first division of the zygote.

(E) Each cell produced by the zygote from the first division is different.

86. Major changes in the mechanism of external respiration have occurred during the evolution of animals, which have tended to optimize the rate of diffusion. Natural selection favored which of the following changes?

    (A) an increase of the surface area over which diffusion occurs

    (B) a decrease in the distance across which diffusion takes place

    (C) an increase in the concentration difference between the internal and external environment.

    (D) choices A and B

    (E) choices A, B, and C

87. What prevents the level of glucose from quickly rising to a high level if you introduce a large amount of glucose into your body?

    (A) The enzyme glucase is secreted by the liver to break glucose down into water and carbon dioxide.

    (B) The excess glucose is removed from the blood by the nephron and excreted by the body.

    (C) Excess glucose is transported to the liver and is stored temporarily as glycogen.

    (D) The glucose is actively transported into muscle cells, where ATP is made and stored in the mitochondria for future use.

    (E) Nothing prevents the quick rise in glucose levels.

88. Pigeons in the park, when they are young, will flee from passing pedestrians. Over time, they learn that the people are not a threat and will not approach or harm them. Consequently, they stop running from people who get close to them. They have learned not to respond to the stimulus of a person approaching them. This behavior is an example of

    (A) Pavlovian conditioning

    (B) classical conditioning

    (C) operant conditioning

    (D) sensitization

    (E) habituation

89. All enzymes

    (A) are proteins

    (B) lower the activation energy of reactions

    (C) operate at the same optimum pH

    (D) can be identified because their names end in '-ase'

    (E) require a coenzyme to work effectively

90. Cell surface markers are made of which of the following to allow 'self' recognition?

    (A) spectrin

    (B) clathrins

    (C) phospholipids

    (D) glycolipids

    (E) glycoproteins

91. Homologous pairs of chromosomes are restored as a result of

(A) fertilization

(B) regeneration

(C) differentiation

(D) germination

(E) gametogenesis

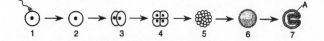

92. During a pregnancy, identical twins could result from the

(A) repetition of the process in stage 1

(B) implantation of the structure in stage 6

(C) separation of the cells in stage 3

(D) differentiation of cells in stage 7

(E) union of two stage 2 cells

93. Of the following, which is NOT a modified stem?

(A) Stipule

(B) Corm

(C) Tuber

(D) Thorn

(E) Rhizome

Question 94 refers to the frog dissection represented in the diagram below.

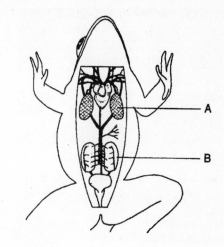

94. Which statement best describes the relationship of structure A to structure B?

(A) A is medial to B

(B) A is dorsal to B

(C) B is lateral to A

(D) B is posterior to A

(E) B is anterior to A

95. The exchange of segments of chromatids during synapsis is known as

(A) cross-linkage

(B) sex-linkage

(C) gene linkage

(D) crossing over

(E) nondisjunction

96. Which of the following is **NOT** true of the members of phylum Chordata?

(A) All are deuterostome coelomates.

(B) All have a backbone.

(C) All have pharyngeal slits at some point in development.

(D) All have a notochord at some point in development.

(E) All have a dorsal hollow nerve cord at some point in development.

97. In an endergonic reaction, the products of a reaction contain

(A) more energy than the reactants, and energy is released

(B) more energy than the reactants, and energy must be supplied

(C) less energy than the reactants, and energy is released

(D) less energy than the reactants, and energy must be supplied

(E) the same amount of energy than the reactants, but due to the presence of an enzyme, energy is released

98. The differential reproductive success of individuals who possess traits that allow them to successfully interact with their environment is

(A) use and disuse

(B) natural selection

(C) artificial selection

(D) reproductive isolation

(E) reproductive selection

99. An example of a positive feedback loop would be

(A) a rise in body temperature causing a dilation of blood vessels in the skin

(B) the patellar reflex

(C) the production of epithelial cells to heal a cut

(D) blood clotting

(E) constriction of the iris of the eye in bright light

100. The secretion of which of the following stops stomach contractions and prevents additional chyme from entering the duodenum?

(A) cholecystokinin

(B) gastrin

(C) secretin

(D) gastric inhibitory peptide

(E) chymotrypsin

101. Aposematic coloration serves to

(A) attract females to males

(B) camouflage individuals

(C) identify the mother to its offspring

(D) keep potential predators away from poisonous or otherwise dangerous prey

(E) confuses predators with the similarity to other poisonous organisms

102. Which process is most similar to the process of cloning?

(A) fertilization

(B) vegetative propagation

(C) meiosis

(D) oogenesis

(E) gamete formation

103. Social insect colonies are composed of highly integrated groups called

    (A) unitaries
    (B) families
    (C) castes
    (D) workers
    (E) colonies

104. What will happen if a base sequence of a strand of DNA is changed from A–T–G to A–T–C?

    (A) The mRNA will be changed from U–A–C to U–A–G
    (B) The mRNA will be changed from T–U–C to T–U–G
    (C) The tRNA will be changed from U–A–C to T–A–C
    (D) The tRNA will be changed from C–A–U to C–A–C
    (E) The tRNA will be changed from A–T–G to A–T–C

105. While the electron transport chain is the more efficient way to produce ATP by eukaryotes, they also produce ATP by substrate-level phosphorylation. This involves

    (A) bonding a free phosphate from the cytoplasm to ADP
    (B) cleaving a phosphate from an inorganic metallic-phosphate compound
    (C) the transfer of a phosphate group from NADP
    (D) cycling a phosphate group from the output of the oxidation-reduction reactions during the electron transport chain
    (E) transferring a phosphate group to ADP from a phosphate-bearing intermediate

106. Gene frequencies in a population would most likely change due to

    (A) random mating
    (B) a stable environment
    (C) a large population
    (D) an increase in birth rate
    (E) mutations

107. During translation, what must happen prior to the ribosome moving to the next codon on the mRNA molecule?

    (A) formation of the initiation complex
    (B) formation of a peptide bond
    (C) the smaller ribosomal subunit must attach to the initiation codon
    (D) an aminoacyl-tRNA attaches to the P site
    (E) a tRNA-peptide chain attaches to the A site

108. The basic photosynthetic unit in a chloroplast is the

    (A) stroma
    (B) stoma
    (C) thylakoid
    (D) granum
    (E) chlorophyll

109. Reciprocal exchange is a process that occurs

    (A) when pieces of homologous chromosomes are exchanged

    (B) when equal number of ions enter and leave a cell

    (C) in the colon of invertebrates to achieve water balance

    (D) when plasmids are cut out or inserted into a bacterial cell's main genome

    (E) to balance the salt content of the plasma of the blood

110. Impulses conducted along the transverse tubules

    (A) stimulate the sodium-potassium pump

    (B) inhibit the binding of Ca++ to troponin

    (C) allow active transport of Ca++ into the sarcoplamsic reticulum

    (D) stimulate the release of Ca++ ions into the cytoplasm from the sarcoplasmic reticulum

    (E) allow cross-bridges to form between actin and myelin

111. Which of the following groups is a deuterostome?

    (A) tunicates

    (B) annelids

    (C) platyhelminthes

    (D) arachnida

    (E) crustacea

112. When a deletion interrupts a codon

    (A) the downstream portion of the gene is transcribed out of register

    (B) transcription ceases

    (C) RNA polymerase moves through cytoplasm to find the missing amino acid

    (D) the defective codon is skipped and proper transcription resumes with next intact codon

    (E) DNA polymerase excises the other 2 amino acids of the codon during transcription

113. Of the following functional chemical groups, which is the only one that contains nitrogen?

    (A) carboxyl

    (B) animo

    (C) methyl

    (D) hydroxyl

    (E) sulfhydryl

114. DNA fragments from different genomes that were cut by the same restriction endonuclease can be spliced together because

    (A) DNA polymerases can reunite cut pieces of DNA

    (B) all restriction endonucleases cut DNA into short fragments

    (C) scientists have developed techniques that allow cut pieces of DNA to be rejoined, regardless of how they have been cut

    (D) all DNA is made of the same four nucleotides, with sequencing the only difference

    (E) the restriction endonuclease cleaves in a specific place, and the cleaved ends are complementary

115. Which of the following statements is correct about the graph shown below?

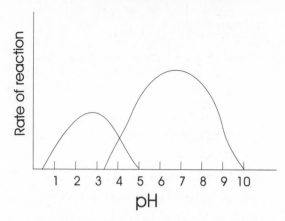

(A) The optimum pH of each enzyme overlaps.

(B) Neither enzyme will work in the presence of the other enzyme.

(C) At a pH of 4.5, the activity of each enzyme cancels the other one.

(D) The optimal pH of both enzymes is acidic.

(E) Enzyme A is stronger than enzyme B.

116. Chylomicrons are

(A) triglyerides that combine with proteins into a water soluble particle, which is absorbed into the lymphatic system

(B) the products of fat digestion from the bile salts

(C) microscopic organisms that belong to the phylum Sarcodina

(D) a unit of measurement that equals 0.10 micron

(E) the small particles found in chyme which have been formed as the result of gastric juice activity

117. Which of the following situations will not stimulate the secretion of salivary juices?

(A) Thinking about dinner on your ride home from work

(B) Coming to class and finding out you have a surprise quiz

(C) Chewing a cracker

(D) The odor of dinner cooking when you walk into the kitchen

(E) The presence of a noxious substance in your stomach

Questions 118 to 120 are based on the chemical reaction shown below.

$$\underset{X}{C_6H_{12}O_6} + \underset{Y}{C_6H_{12}O_6} + (C_6H_{12}O_6)_n ————> \text{polysaccharide} + Z$$

118. Letter Y most likely represents

(A) a neurotransmitter

(B) a hormone

(C) a lipid

(D) an enzyme

(E) DNA polymerase

119. Letter Z most likely represents molecules of

(A) water

(B) plant hormones

(C) glycogen

(D) nucleic acids

(E) chlorophyll

120. If this reaction occurs in an organism that requires sunlight to produce substance X, the organism must be

(A) a heterotroph

(B) an annelid

(C) a chemotroph

(D) an autotroph

(E) a fungus

Peterson's AP Success: Biology

SECTION II

Time–1 Hour and 30 Minutes

Directions: Answer all questions. Number your answers as the questions are numbered below. Answers must be in essay form. Outline form is NOT acceptable. Labeled diagrams may be used to supplement discussion, but in no case will a diagram alone suffice. It is important that you read each question completely before you begin to write.

1. In what ways do organisms benefit from association with other organisms? Give examples to support your discussion.

2. What are the pieces of evidence for evolution? Give examples to support each.

3. As organisms became complex from their unicellular ancestors, the need for communication between parts of the organism became necessary. What types of communication occur between parts of multicellular organisms?

4. Plant and animal cells are very similar; however, there are several differences. Using human cells as your reference for an animal cell and angiosperm cells as your reference for plants, choose any three of the four areas below and discuss the differences between plant and animal cells.
   a. cell division
   b. organelles and other structures present
   c. functions performed
   d. motility of cells

ANSWERS AND EXPLANATIONS

SECTION I

1. **The correct answer is (C).** Each fatty acid combining with the glycerol releases one water molecule. Since there are three fatty acids in a tricglyceride, three molecules of water are released.

2. **The correct answer is (B).** Both the cell wall and chloroplasts are large enough to be seen under low power of the light microscope. The other choices are all too small to be seen under this magnification.

3. **The correct answer is (C).** Vessel elements are joined end to end and have pores. Tracheids do not have any cytoplasm present and conduct only water. Pores are present in vessel elements, while tracheids have pits. The two roles of tracheids in gymnosperms are water transport and structural support.

4. **The correct answer is (B).** Meissner's corpuscles sense touch and are found in the dermis, but they are closer to the surface of the body than Pacinian corpuscles, which sense pressure. Only free nerve endings are wrapped around the hair follicle. Meissner's corpuscles are not necessarily associated with free nerve endings.

5. **The correct answer is (E).** Ribosomes synthesize proteins. Because most enzymes are proteins, their production would cease. Other cellular functions would eventually shut down following a cessation in protein synthesis.

6. **The correct answer is (B).** This is regeneration as exemplified by planaria. Choice (A) is vegetative reproduction involving stolons, or runners, in strawberries. Choice (C) is binary fission as found in bacteria. Choice (D) is release of spores from a fungal sporangium and

choice (E) is a type of sexual reproduction called conjugation found in bacteria.

7. **The correct answer is (C).** The scapula is part of the pectoral girdle, which is part of the appendicular skeleton. The others are parts of the axial skeleton.

8. **The correct answer is (D).** The red color in rhodophyta is due to phycobilins. Some invertebrates use hemocyanin as their respiratory pigment, which imparts a bluish color to the hemolymph. The other two answers are nonsense.

9. **The correct answer is (B).** Glucose is split onto pyruvate (glycolysis) in the cytoplasm. From there, the products of glycolysis travel to the mitochondria for cellular respiration. Cellular respiration occurs in the mitochondria, while the electron transport chain occurs in the inner wall of the mitochondria. The ribosomes, choice (A), are the site for protein synthesis.

10. **The correct answer is (C).** The ampullae of Lorenzini are electrore-ceptors used by elasmobranchs (e.g., sharks and rays) to detect prey.

11. **The correct answer is (D).** These are the three primary meristems. The lateral meristem is secondary growth. Cork cambium is a type of lateral meristem.

12. **The correct answer is (D).** Guanine, cytosine, and adenine are present in both DNA and RNA, while thymine is present only in DNA. Only uracil is present in RNA only.

13. **The correct answer is (E).** While a frog has lungs, the greatest amount of gas exchange occurs through its skin. An amoeba exchanges gases through its cell membrane, an earthworm through its skin, a grasshopper through tracheal tubes, and a human has well-developed lungs for gas exchange.

14. **The correct answer is (A).** The transfer of pollen to the stigma of the same flower is self-pollination. Cross-pollination involves two flowers. Ovulation is the release of an ovum from an ovary, and spermatogenesis is the development of sperm. Differentiation applies to the changes that an unspecialized cell undergoes to become more specialized.

15. **The correct answer is (E).** This type of reflex is termed monosynap-tic.

16. **The correct answer is (A).** The antidiuretic hormone is produced in the hypothalamus and released from the posterior lobe of the pituitary. All of the other choices are produced in the anterior lobe.

17. **The correct answer is (B).** This is the point near the end of G_1, just before beginning the S phase, where the decision is made on whether the cell will divide or not. If the DNA is damaged, this is the point where DNA replication is stopped by the tumor-supressor gene.

18. **The correct answer is (A).** Evidence shows that production of ethylene increases when a plant is subjected to stress. All of the situations will stress the plant except the spraying of insecticides.

19. **The correct answer is (A).** Tight junctions are found between epithelial cells. The cells are in direct contact with each other. These junctions prevent leakage of materials between the cells. Gap junctions, also found in animal cells, consist of pores that allow for passage of large molecules between cells. Plasmodesmata are found in plants, while

aherens junctions connect the actin filaments of one cell to another and desmosomes connect the cytoskeletons of neighboring cells.

20. **The correct answer is (C).** Peas do exhibit sex-linked characteristics. They were not used by Mendel. The other four characteristics are found in peas.

21. **The correct answer is (B).** While some members of the phylum Rhynochocoela have a complete circulatory system, the only system that is present in all acoelomates is a digestive system.

22. **The correct answer is (B).** Contractile vacuoles remove water from protists, such as paramecium. The other choices are all mechanisms of moving substances into cells. Phagocytosis moves particles into a cell, while pinocytosis moves liquids into cells. Porin channels allow the passage of water into cells. Receptor-mediated endocytosis is a mechanism for moving macromolecules into cells at clathrin-coated pits.

23. **The correct answer is (E).** Only the crustaceans have these structures. The uropod is the flattened, paddle-like appendage at the posterior end of the crustacean. The carapace is the dorsal shield arising from the head. The swimmerets are lateral appendages of the abdomen that are used for swimming and reproduction.

24. **The correct answer is (A).** Gastrulation involves division of the one cell-layered blastula to form the gastrula, which is made of three layers. Blastulation is the formation of a blastula, which is a hollow ball of one cell layer from the zygote. Cleavage is also the formation of the blatula. Meiosis is the process where the diploid chromosome number is reduced to the haploid condition. Nondisjunction is a chromosome mutation where, during meiosis, the homologous chromosomes do not separate, resulting in a gamete with two of the same chromosomes.

25. **The correct answer is (D).** This is the definition of oxidation. Choice (A): reduction involves an atom or molecule gaining an electron. The other terms involve energy but not specifically the loss of an electron during a reaction.

26. **The correct answer is (E).** This is the chamber that pumps the systemic blood, and the pressure in this chamber is the highest.

27. **The correct answer is (D).** Haversian canals are narrow channels running lengthwise through bone tissue and containing blood vessels and nerves. The hardened matrix of bone is made of crystals of calcium phosphate. Nails and hooves are made of protein, while cartilage does not contain canals and is made of a glycoprotein. Teeth contain calcium phosphate-fluoride crystals in the enamel, but no canals.

28. **The correct answer is (A).** The fatty acids that result from the breakdown of chylomicrons may be taken up by the cells of the body.

29. **The correct answer is (A).** The cardiovascular center in the brain, the medulla oblongata, receives impulses from the baroreceptors. When the pressure in the aorta and carotid artery decreases, the frequency of these impulses to the medulla decrease, causing stimulation of the sympathetic nervous system. The stimulation increases the rate and force of the contractions by the ventricles.

30. **The correct answer is (E).** Thigmotropism is directional growth in response to touch. Photoperiodism relates to growth in response to light. Gravitropism relates to response to gravity. Both cytokinins and gibberellins are plant hormones.

31. **The correct answer is (A).** Smooth muscles appear smooth because their actin and myosin filaments are not arranged in regular sarcomeres.

32. **The correct answer is (C).** The medulla, also known as the medulla oblongata, is the control site for involuntary activities, such as breathing and heartbeat.

33. **The correct answer is (D).** Bicarbonate, glucose, amino acids, NaCl, and water are all reabsorbed from the proximal tubule. Water continues to be reabsorbed from the descending limb. NaCl is again reabsorbed from the ascending limb and from the distal tubule. Water is again recovered from the collecting duct.

34. **The correct answer is (C).** Flowers are the sexual reproductive structures of plants, so a plant that does not produce flowers cannot reproduce sexually.

35. **The correct answer is (A).** This is the major advantage of a closed circulatory system in a multicellular organism. While choices (B), (C), and (D) are true statements, they are not major advantages of a closed circulatory system, and their results can be accomplished in other ways. Choice (E) is not a true statement.

36. **The correct answer is (A).** Being ions, both sodium and potassium cannot pass through the cell membrane directly and must move through a gated channel, which is the sodium-potassium pump.

37. **The correct answer is (A).** Interferons and natural killer cells are both parts of the nonspecific defense strategy.

38. **The correct answer is (A).** The Islets of Langerhans secrete insulin, which is a hormone. The other four choices are cells that secrete digestive enzymes.

39. **The correct answer is (C).** The testes need an environment that is 2 to 4 degrees cooler than the interior of the abdominal cavity in order to produce viable sperm. They form in the abdominal cavity but descend into the scrotum prior to birth.

40. **The correct answer is (C).** In artificial selection, the choice of which characteristics to breed is selected by a breeder, such as a farmer or cattleman.

41. **The correct answer is (E).** This is the definition of an allosteric inhibitor. A competitive inhibitor attaches to the active site, choice (A). Choices (B) and (C) don't refer to noncompetitive inhibition, and choice (D) makes no sense.

42. **The correct answer is (B).** Emulsification of fats occurs in the small intestines through the action of bile, which is produced by the liver.

43. **The correct answer is (B).** It was necessary first to be able to store energy in the bonds of molecules, and ATP is the energy carrier in all organisms today. Then catalysis had to occur in organized sequences. Photosynthesis first began using hydrogen sulfide because the environment was anaerobic. In order for proteins and nucleic acids to be synthesized, it was necessary to obtain nitrogen atoms from nitrogen gas (N_2), as nitrogen is a component of both of these molecules. When water was used for photosynthesis, rather than hydrogen sulfide, oxygen gas was released. Finally, organisms evolved that could subsist entirely from breaking down organic molecules to gain energy.

44. **The correct answer is (D).** These locations contain cuboidal epithelium. The heart walls contain cardiac muscle, the knees and other joints contain cartilage, the surface of the skin is covered with squamous epithelium, and the tendons and sheaths around voluntary muscles are made of dense connective tissue.

45. **The correct answer is (C).** The probability of each child having a characteristic is independent of the characteristics of each prior child. Since the mother is a carrier for the trait, there is a 50 percent probability that she will pass that trait to her next son. Since the trait is sex-linked, there is no compensating gene on the Y chromosome given by the father, so the probability of the male child having the condition is 50 percent.

46. **The correct answer is (C).** The telson is a tail spine and is not considered an appendage. The other four choices are all appendages.

47. **The correct answer is (C).** Cytosine is complementary to guanine, uracil is complementary to adenine, and adenine is complementary to uracil. Thymine is not present in RNA.

48. **The correct answer is (B).** Analogous structures are those that serve the same function but have different embryological origins. The butterfly wing is analogous to all of the other appendages, which are homologous to each other. They have a similar structure and the same embryological origin but a different function.

49. **The correct answer is (E).** This is the definition of a favorable adaptation.

50. **The correct answer is (A).** This is the definition of adhesion. Cohesion is the attraction of water molecules to other water molecules, and it causes the surface tension of water. The bonds between water molecules are due to hydrogen bonding. Specific heat is due to hydrogen bonds that absorb heat when they break.

51. **The correct answer is (B).** These two pigments are characteristic of the phylum Chlorophyta, or the green algae.

52. **The correct answer is (D).** This is the major reason that amphibians are tied to life in the water. They must lay their eggs in the water for fertilization to occur and to prevent dessication of the eggs.

53. **The correct answer is (B).** Anemia is an insufficient amount of hemoglobin or an inadequate number of red blood cells.

54. **The correct answer is (D).** Some flies have a natural immunity to insecticides, based on their genes.

55. **The correct answer is (A).** Utilization of sugar and oxygen to make ATP, carbon dioxide, and water is found in the mitochondria. The reverse process occurs in the chloroplast.

56. **The correct answer is (E).** Humans do not make aromatic amino acids but need to get them from their diet, so glyphosate has no effect.

57. **The correct answer is (B).** The pancreas contains the Islets of Langerhans, which secrete insulin. Insulin is responsible for the metabolism of sugar. Dogs who lack a pancreas will not be able to break down sugars, so it will be excreted in the urine. This sugar in the urine attracts the flies.

58. **The correct answer is (E).** The presence of air sacs serves to store supplemental air for the large demands placed upon the flight muscles of birds.

59. **The correct answer is (B).** Leukocytes are formed from the same stem cells as erythrocytes but differentiate differently.

60. **The correct answer is (D).** The pituitary secretes luteinizing hormone (LH) and the follicle stimulating hormone (FSH), and the ovaries secrete estradiol and progesterone, both of which control the cyclic nature of the menstrual cycle.

61. **The correct answer is (A).** The notochord develops from the mesoderm. Somites and neural groove develop after the notochord. The primitive streak is made of ectoderm.

62. **The correct answer is (B).** These are the guard cells, which lose turgor when water flows out of them, allowing them to close the stoma.

63. **The correct answer is (B).** The Haversian canals are the channels containing blood vessels and nerves, and the lamellae are the rings of bone tissue.

64. **The correct answer is (A).** This is the definition of epistasis.

65. **The correct answer is (E).** These are secondary sex characteristics in males.

66. **The correct answer is (A).** The sponge digests its food intracellularly.

67. **The correct answer is (E).** DNA polymerase adds nucleotides to the 3' end of the new strand. As a result, all DNA, including Ozaki fragments, are synthesized in the 5' to 3' direction. Since the two template strands are antiparallel, the two new strands are made in opposite directions.

68. **The correct answer is (C).** Since the sperm must swim to the egg once they have been deposited into the female reproductive tract, the fluid medium serves as a transport medium for the sperm.

69. **The correct answer is (D).** This is movement of water between the cells of the plant, hence it is intercellular transport.

70. **The correct answer is (A).** This is the correct order of the activities in protein synthesis.

71. **The correct answer is (B).** Due to the electrical nature of polar molecules, water molecules will gather closely around them, and the charged portion of the water molecule will be attracted to the charged region of that molecule. The oxygen in water is attracted to positively charged molecules; hydrogen ions in water are attracted to negatively charged molecules.

72. **The correct answer is (E).** Some animals, including the Himalayan hare, have the ability to change their phenotype in response to environmental influences, such as cold weather.

73. **The correct answer is (A).** Barr bodies are inactivated X chromosomes in females.

74. **The correct answer is (B).** The nematodes were the first to evolve a body cavity. They have a pseudocoelom. Molluscs and annelids have a true coelom, while cnidarians and flatworms, the platyhelminthes, are acoelomate. As such, they lack a body cavity.

75. **The correct answer is (B).** Amylase is secreted by the salivary glands into the mouth. Pepsin is secreted into the stomach. Sucrase and trypsin are both secreted into the small intestine.

76. **The correct answer is (A).** This is the yolk sac, which supplies food for the developing organism.

77. **The correct answer is (E).** This is the amniotic sac, which serves as a cushion for the developing embryo.

78. **The correct answer is (B).** The chorion is the outermost membrane. It aids in the exchange of gases between the embryo and the environment.

79. **The correct answer is (B).** These are homologous structures, having a common ancestor even though their functions are different.

80. **The correct answer is (E).** This is the definition of punctuated equilibrium. Gradualism is a theory that promotes slow, gradual evolutionary change.

81. **The correct answer is (E).** RNA contains uracil and is complementary to the adenine in DNA. Guanine and cytosine are complementary. Thymine is present in DNA but not RNA.

82. **The correct answer is (C).** G_o is a nondividing stage. Most cells of the adult human body are in G_o. Some cells can return to cell division from G_o in response to environmental cues.

83. **The correct answer is (E).** The cytoskeleton is made of these three components. Myosin is a protein filament that is found in muscle tissue and is not part of the cytoskeleton.

84. **The correct answer is (B).** The condition pictured is trisomy-21 or Down's Syndrome. There was nondisjunction of chromosome 21, resulting in the zygote getting three number 21 chromosomes.

85. **The correct answer is (D).** This is a characteristic of deuterostomes. Choices (A), (B), and (C) are protostome characteristics. Choice (E) is not necessarily a characteristic of either.

86. **The correct answer is (E).** All three of these events were favored by natural selection. By increasing the rate of diffusion of gases, these organisms were favored because they were better able to gain oxygen for muscles for fighting and/or fleeing, as was appropriate.

87. **The correct answer is (C).** When glucose levels exceed normal values, the excess is transported through the blood to the liver, where it is temporarily stored as the polysaccharide glycogen.

88. **The correct answer is (E).** This is habituation, which means learning not to respond to a meaningless stimulus. Pavlovian conditioning involves paired stimuli causing a response from a stimulus that has nothing to do with the response. Classical conditioning is a general term that includes Pavlovian conditioning. In operant conditioning, an animal learns to associate a behavioral response with a reward or punishment. In sensitization, which is a type of nonassociative learning, there is an increased response to a stimulus.

89. **The correct answer is (B).** Lowering the activation energy allows a reaction to occur more quickly. Enzymes do this. Until recently, choice (A) was considered correct, but there is strong evidence that some RNA molecules, referred to as ribozymes, serve as enzymes. Choices (C), (D), and (E) are all incorrect statements. Many enzyme names end in "-ase"—but not all.

90. **The correct answer is (E).** Glycolipids are for tissue recognition, as in the ABO blood group markers. Spectrin, found in red blood cells,

determines the shape of the cell. Clathrins are on the surface membranes of cells and facilitate binding of molecules to the membrane. Phospholipids comprise the actual cell membrane itself.

91. **The correct answer is (A).** Each gamete, the sperm and egg, contribute one of the homologous pair of chromosomes when they unite during fertilization.

92. **The correct answer is (C).** After the zygote has divided once, the separation and subsequent division and differentiation of these two cells results in identical twins—two individuals with the same genes. A repetition of the process in stage 1 would result in fraternal twins, because two different eggs are fertilized by two different sperm.

93. **The correct answer is (A).** Stipules are leaflike, or if reduced, spinelike or glandlike appendages at the base of a petiole. Corms are thick, fleshy underground stems modified for storage of food. Tubers are underground stems with fleshy scale leaves. Thorns are modified branches, as found in plants, such as the hawthorne. Rhizomes are underground stems.

94. **The correct answer is (D).** Posterior is toward the tail end and anterior is toward the head end, so B is posterior to A.

95. **The correct answer is (D).** When the homologous chromosomes come together during synapsis, they are in close contact with each other, and this is the time when parts of the chromosomes may exchange segments.

96. **The correct answer is (B).** Only the vertebrates, a subset of the Chordates, have a backbone.

97. **The correct answer is (B).** This is the definition of the endergonic reaction. Choice (C) is the definition of an exergonic reaction. Choices (A), (D), and (E) make no sense.

98. **The correct answer is (B).** This is natural selection. Favorable characteristics allow some individuals in a population to successfully adapt to changes. Use and disuse was the Lamarckian theory, which was shown to be incorrect. Artificial selection is done by breeders of animals and plants. Reproductive isolation involves an event that forms a barrier to breeding between two isolated groups. Reproductive selection is a ficticious term.

99. **The correct answer is (D).** In a positive feedback loop, changes in the body cause a variable to move successively away from its set point. The formation of a blood-clotting factor activates another in a cascade that leads to the formation of a clot. None of the other examples involve a series of events that move a variable continually farther from its homeostatic set point.

100. **The correct answer is (D).** Cholecystokinin stimulates gall bladder secretion and secretion of enzymes by the pancreas. Gastrin is a hormone secreted by the stomach to stimulate the secretion of hydrochloric acid and pepsin by the stomach. Secretin is secreted by the duodenum and stimulates secretion of bicarbonate by the pancreas, while chymotrypsin is an enzyme secreted by the pancreas for protein digestion.

101. **The correct answer is (D).** Many toxic or unpalatable animals are brightly colored, which serves to warn potential predators to stay away.

102. **The correct answer is (B).** This means making new individuals asexually in plants, while cloning is associated with creating new individuals from cells in animals.

103. **The correct answer is (C).** Kin are family members. The family is a social group of higher animals or a taxon. Workers are one caste of a social society, and colonies are groups of individuals, such as bacteria, or breeding colonies of birds.

104. **The correct answer is (A).** Since uracil is present in RNA instead of thymine, which is present in DNA, the mRNA which originally was U–A–C becomes U–A–G.

105. **The correct answer is (E).** This is the definition of substrate-level phosphorylation. The "phosphate-bearing intermediate" is the substrate.

106. **The correct answer is (E).** Mutations are a major cause of changing gene frequencies. The other four choices all contribute to the stability of gene frequencies.

107. **The correct answer is (B).** A peptide bond must be formed before the ribosome moves to read the next codon on the mRNA.

108. **The correct answer is (C).** This flattened sac, groups of which make up the granum, is the basic photosynthetic unit and is the site of the light reactions. A stoma is an opening into a leaf or stem. The stroma is the matrix containing the grana and contains the Calvin cycle enzymes of the light-independent reactins. Chlorophyll is the green pigment found in chloroplasts.

109. **The correct answer is (D).** This is a source of genetic variability in bacteria.

110. **The correct answer is (D).** Ca++ ions are stored in the sarcoplasmic reticulum, a specialized type of endoplasmic reticulum that surrounds the myofibrils of muscle cells. When an impulse is conducted along the transverse tubules, the sarcoplasmic reticulum is stimulated to release the Ca++ ions.

111. **The correct answer is (A).** Deuterostomes have their anus formed from the blastopore. The other four organisms are protostomes, which have their mouth formed from the blastopore.

112. **The correct answer is (A).** This is the result of what a deletion does, since it removes part of the message, and the remaining part of the message on the DNA is transcribed into mRNA without this part. When the tRNA brings the amino acid to the ribosome to build the protein, it brings the incorrect amino acid from the point of deletion on.

113. **The correct answer is (B).** The amino group is NH_3. Carboxyl is COOH, methyl is CH_3, hydroxyl is OH, and sulfhydryl is SH.

114. **The correct answer is (E).** Restriction enzymes cut DNA at characteristic sequences. Since the cleaved ends are complementary, they can pair with each other into chains.

115. **The correct answer is (D).** The peak of curve of activity of an enzyme is its optimal activity. Enzyme A has an optimal point of about 3, while enzyme B has an optimum of about 6.5. Both of these points are acidic. While both enzymes will function in the pH range of 4 to 5, this is not the optimal point of either. There is no reason to suspect

Peterson's AP Success: Biology

that these enzymes will cancel each other out, and the fact that they work in different pH ranges has nothing to do with their strength.

116. **The correct answer is (A).** Glycerol and fatty acids are absorbed from the intestine into epithelial cells. Here, they are combined to form fats, mixed with cholesterol, coated with proteins, and released to the lymphatic system as chylomicrons.

117. **The correct answer is (B).** All of the other situations will stimulate secretion of salivary juices: thinking about food, the presence of food in the mouth, smelling food, or the presence of a noxious substance in the stomach. Coming to class to find you have a surprise quiz will most likely stimulate the secretion of adrenaline.

118. **The correct answer is (D).** This is a synthesis reaction to form a polysaccharide from monosaccharides. It would be catalyzed by an enzyme. Enzymes are written on the arrow because they are neither reactants nor products in reactions.

119. **The correct answer is (A).** Synthesis reactions combine smaller molecules into a larger molecule, with the formation of water as a by-product.

120. **The correct answer is (A).** Substance X is a monosaccharide or simple sugar, such as glucose. Organisms that require sunlight to produce simple sugars are autotrophs. A heterotroph needs to eat to gain its food. An annelid is a segmented worm. Chemotrophs are organisms that utilize inorganic molecules to make food, and a fungus is a member of the kingdom Fungi, which is heterotrophic.

SECTION II

1. Students receive 1 point for each of the following concepts and 1 point for each correct example.

   Mutualism is two organisms living together, and both organisms benefit from the association.

   Commensalism is when two organisms live together, and only one of the organisms benefits, while the other is unaffected.

   Parasitism is when two organisms live together, and one organism benefits, while the other is harmed.

   Predation is when one organism eats another.

   Mimicry is when one organism copies some aspect of another organism for the purpose of protection.

   Camouflage is when one organism uses its surroundings, frequently plants, to conceal itself.

2. Students receive 1 point for each piece of evidence and 1 point for each example.

   Fossil record is the location of fossils within a rock. Older fossils are found deeper than newer fossils. When examining the differences in fossils, one can see the differences that have occurred over time.

   Molecular record. Those organisms, which have evolutionary ties, have more similarities in the molecular structure of their DNA and proteins than do more distantly related organisms. The protein cytochrome c is frequently used for comparison.

Development of an organism from a zygote to an adult shows similarities between vertebrates. Tails, limb buds, and pharyngeal gill slits all suggest that humans are closely related to less advanced vertebrates.

Vestigial structures are those structures which no longer serve a purpose for an organism. Examples of these in humans include ear muscles and an appendix. These structures have functions in simpler vertebrates, suggesting that there is an evolutionary relationship between humans and these organisms.

Convergent evolution is the evolution of organisms along similar lines with no common ancestor. The presence of the marsupials in Australia and their resemblance to the placental mammals of the remaining world suggest that a parallel evolution has occurred with similar selection.

Patterns of distribution, particularly on islands close to a mainland, show that the organisms on these islands resemble, but are not identical to, organisms on the mainland. There are some differences, but the similarities are strong enough to suggest that the island organisms have evolved from their mainland counterparts.

3. Students receive 2 points for each concept answered.

A circulatory system connects parts of a multicellular body to allow the movement of food, oxygen, wastes, and carbon dioxide, plus hormones.

A nervous system allows for the reception of information from parts away from the central processing area to be picked up and transferred to the central nervous system, where it is processed and decisions are made. This information is from both external and internal sources.

Receptors include the five senses, as well as internal receptors, for information such as blood pressure, peristalsis, urine in the bladder, digestion, and respiration.

An endocrine system contains organs that release chemical messengers into the blood to be carried to other parts of the body, where they effect some change. Common examples are the hormone thyroxine released from the thyroid, which affects cell metabolism; adrenaline, released from the adrenal medulla, which increases the heartbeat, blood pressure, and supply of blood to the heart muscle; and insulin produced by the pancreas, which stimulates the liver, skeletal muscle, and adipose tissue to take up and use glucose.

4. Students receive 2 points for each of the following concepts, with no more than 4 points awarded in any of the four parts (a through d). The question is worth 10 points total.

Animal cells have centrioles that serve to organize spindle fibers during mitosis; these are lacking in plant cells.
Animal cells divide by forming a cleavage furrow, an indentation in the cell membrane surrounding the entire cell; plants form a cell plate, which is laid down at the equator of the cell at the end of telophase.
Animal cells elongate during cell division, prior to the formation of the cleavage furrow subsequent division, while plant cells do not increase in length prior to the formation of the cell plate.

Animal cells have centrioles, flagella, lysosomes, and cilia.

Plant cells may contain chloroplasts and have cell walls and a central vacuole.

Plant cells perform photosynthesis, the conversion of water and carbon dioxide into sugar and oxygen using sunlight. Animals do not; they obtain their energy by eating nutrient materials.

Animal cells digest materials in lysosomes, while plants do not.

Plant cells maintain turgor by keeping a central vacuole filled with water. Animal cells lack a central vacuole.

Animal cells store carbohydrates as starches, while plant cells store starches.

Plants lack motile cells, while animals have motile cells in the male reproductive cells only.

PRACTICE TEST 3

AP BIOLOGY

SECTION I
Time–1 Hour and 30 Minutes
Number of Questions–120
Percent of Total Grade–60

Directions: Each of the questions or incomplete statements below is followed by five suggested answers. Select the one that is best in each case.

1. Water can move from the roots of the trees to their uppermost branches, and basilisk lizards can run on water for brief periods of time because of surface tension. Surface tension is the property of water caused by

   (A) attraction of a hydrogen atom of one water molecule to the oxygen of another water molecule
   (B) attraction of an oxygen molecule of water to one or more hydrogen atoms of other water molecules
   (C) short-lived hydrogen bonds between water molecules
   (D) cohesion
   (E) all of the above

2. The amount or percentage of water found in human blood is about

   (A) 25 percent
   (B) 50 percent
   (C) 66 percent
   (D) 85 percent
   (E) 92 percent

3. The molecular weight of sucrose is 342 grams. If a student added 171 grams of sucrose to 250 milliliters of water, the approximate molarity would be

   (A) 0.5
   (B) 1.0
   (C) 1.5
   (D) 2.0
   (E) 4.0

4. A solution with a pH of 4 has how many more hydrogen ions than one with a pH of 8?

   (A) 320
   (B) 4000
   (C) 10000
   (D) 1/4000
   (E) 1/10000

Questions 5-7 are based on the following drawing.

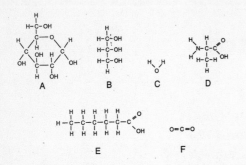

A   B   C   D

E   F

5. Which of the molecules above is the primary energy source for animals?

(A) A

(B) B

(C) F

(D) D

(E) E

6. Lipids are composed of molecules like those in letters

(A) A and E

(B) B and E

(C) A and D

(D) B and D

(E) E and F

7. The letter of the molecule that represents an amino acid is

(A) A

(B) B

(C) F

(D) D

(E) E

Questions 8 and 9 are based on the following table.

Messenger RNA (mRNA) Codes for Selected Amino Acids	
Amino Acid	**mRNA Code**
Leucine	C–C–A
Arginine	C–G–A
Phenylalanine	U–U–U
Valine	G–U–U
Lysine	A–A–A

8. What would be the DNA triplet that would code for the addition of arginine to a polypeptide?

(A) C- G -A

(B) G- C- T

(C) C- C- T

(D) T- A- C

(E) G- C- U

9. Which amino acid would be attached to a polypeptide chain if the amino acid was attached to the anticodon A-A-A?

(A) phenylalanine

(B) lysine

(C) valine

(D) a stop codon

(E) it cannot be determined from the information given

Questions 10 and 11 are based on the following drawing.

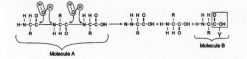

Molecule A   Molecule B

10. The number of peptide bonds in molecule A is

(A) 1

(B) 2

(C) 3

(D) 4

(E) 0

11. The number of amino acids in molecule A is

(A) 1

(B) 2

(C) 3

(D) 4

(E) 0

12. Which expression in the following equation for calculating the free energy in a system reflects changes in the orderliness of the reactants and products?

$$\Delta G = \Delta H + T\Delta S$$

(A) G

(B) ΔH

(C) T

(D) ΔS

(E) $T\Delta S$

Question 13 is based on the following diagrams.

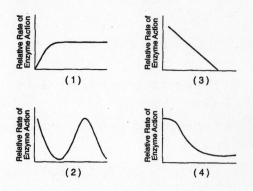

13. Which of the above diagrams represents the relative rate of a catalyzed reaction when the amount of enzyme is held constant and there is increasing amounts of substrate?

(A) 1

(B) 2

(C) 3

(D) 4

(E) none of the above

14. Glucose can be broken down in both alcoholic fermentation and cellular respiration. How many times more efficient is cellular respiration than alcoholic fermentation?

(A) It is not—fungi are the most efficient consumers known to man.

(B) 2 times

(C) 9 times

(D) 18 times

(E) 34 times

15. Consider a reaction in which the free energy of the products is greater than the free energy of the reactants. Which of the following statements is true?

    (A) The reaction is exergonic, and K_{eq} is greater than 1.

    (B) The reaction is endergonic, and K_{eq} is greater than 1.

    (C) The reaction is exergonic, and K_{eq} is less than 1.

    (D) The reaction is endergonic, and K_{eq} is less than 1.

    (E) One needs to calculate K_{eq} for each reaction.

16. The products of glycolysis are

    (A) 2 ATP and 2 NADH

    (B) 2 ATP and 1 NADH

    (C) 2 ATP and 2 NAD$^+$

    (D) 2 ADP and 2 NADH

    (E) 2 ADP and 2 NAD$^+$

17. Complete the following analogy for the production of energy in certain cells. Intermembrane space : _____ as thylakoid compartment :_____ .

    (A) matrix; bundle sheath

    (B) matrix; stroma

    (C) stroma; matrix

    (D) stroma; bundle sheath

    (E) cristae; matrix

18. The protein that does not belong among the others in the respiratory chain is

    (A) NADH reductase

    (B) cytochrome reductase

    (C) NADPH dehydrogenase

    (D) ubiquinone

    (E) cytochrome oxidase

19. Which of the following organelles is out of order from an endomembrane point of view?

    (A) nucleus

    (B) vesicles

    (C) golgi apparatus

    (D) endoplasmic reticulum

    (E) plasma membrane

Question 20 is based on the following diagrams.

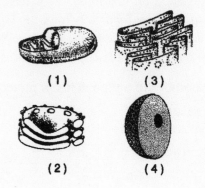

(1)   (3)

(2)   (4)

20. What are the identities of the organelles 1 through 3?

    (A) chloroplast, golgi apparatus, endoplasmic reticulum

    (B) chloroplast, mitochondrion, endoplasmic reticulum

    (C) mitochondrion, golgi apparatus, endoplasmic reticulum

    (D) mitochondrion, endoplasmic reticulum

    (E) endoplasmic reticulum, chloroplast, mitochondrion

21. Intercellular junctions are important in both plants and animals. Which of the following is **NOT** used by animals as junctions?

    (A) tight junctions

    (B) plasmodesmata

    (C) desmosomes

    (D) gap junctions

    (E) communicating junctions

22. Lynn Margulis hypothesized that today's eukaryotic cells are the descendants of an early symbiotic relationship established when an ancestral cell was inhabited by some of the ancestors of today's organelles. Which organelles did she use in her theory?

    (A) nuclei, mitochondria

    (B) nuclei, chloroplasts

    (C) mitochondria, chloroplasts

    (D) mitochondria, vacuoles

    (E) nuclei, golgi apparatus

Question 23 is based on the following diagram.

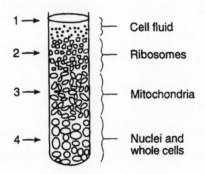

23. Which statement best describes the technique used to obtain these results?

    (A) A compound light microscope was used to show that the organelles in region 1 weigh less than those in region 2.

    (B) An electron microscope was used to show that the organelles in region 3 are the most complex.

    (C) Chromatography was used to determine that the organelles in region 1 are more soluble than those in region 2.

    (D) An ultracentrifuge was used to separate the organelles with varying densities into regions 1 through 4.

    (E) Gel electrophoresis was used to separate the organelles based on size and charge.

24. Plants that use C4 photosynthesis utilize

    (A) rubisco

    (B) phosphoenolpyruvate

    (C) PEP carboxylase

    (D) bundle sheath cells

    (E) all of the above

25. The products of the light reactions, or photophosphorylation, in photosynthesis are

    (A) oxygen and water

    (B) oxygen and ATP

    (C) oxygen, ATP, and NADPH

    (D) water, ATP, and NADPH

    (E) water, ATP, and $NADP^+ + H^+$

Questions 26 and 27 refer to the following diagram. The membrane is semi-permeable, in that it is permeable to NaCl but not glucose.

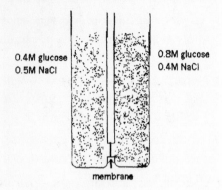

0.4M glucose
0.5M NaCl

0.8M glucose
0.4M NaCl

membrane

26. Choose the statement from below that is true.

(A) Initially, the water potential is higher in the tube on the right, and the water level will rise on the left after several days.

(B) Initially, the water potential is higher in the tube on the left, and the water level will rise on the left after several days.

(C) Initially, the water potential is higher in the tube on the right, and the water level will rise on the right after several days.

(D) Initially, the water potential is higher in the tube on the left, and the water level will rise on the right after several days.

(E) After several days, there will be equal amounts of water in both tubes.

27. Choose the statement from below that is true. After three days, the following conditions would exist:

(A) a decrease in the concentration of NaCl and glucose on the left side

(B) a decrease in the concentration of NaCl on the left side, no change in the glucose level, and an increase in the water level on the right side

(C) no change in the system

(D) a decrease in the concentration of NaCl on the left side and an increase in the water level on the left side

(E) an increase in the concentration of NaCl on the left side and an increase in the water level on the right side

28. Which statement concerning plants that utilize C4 photosynthesis is NOT true?

(A) They open their stomata only at night.

(B) They capture carbon dioxide with PEP carboxylase.

(C) They are more efficient than C3 plants because they transpire less and therefore lose less water.

(D) They use the Calvin cycle to make sugars.

(E) They use bundle sheath cells.

Peterson's AP Success: Biology

29. Which statement concerning the cell cycle is **NOT** true?

   (A) Cyclin is a protein that combines with a cdk (cyclin-dependent kinase) to form a complex known as the maturation promoting factor (MPF).

   (B) The concentration of the molecules cyclin and cyclin-dependent kinases fluctuate during the cell cycle.

   (C) MPF coordinates the stages of mitosis by phosphorylating certain proteins.

   (D) S is the stage of the cell cycle that stands for synthesis.

   (E) The shortest part of the cell cycle is mitosis, not any of the growth phases.

30. Consider a plant cell that has 14 chromosomes and is undergoing prophase in mitosis. At this time there are _____ chromatids and _____ centromeres.

   (A) 14, 14
   (B) 14, 28
   (C) 0, 14
   (D) 0, 28
   (E) 28, 14

31. During meiosis in the human male, the diploid cell gives rise to _____ haploid cells and _____ chromosomes in each cell.

   (A) 2, 46
   (B) 2, 23
   (C) 4, 23
   (D) 4, 46
   (E) 8, 92

Question 32 is based on the following diagram.

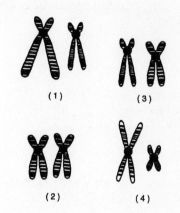

(1)   (3)

(2)   (4)

32. Which represent a homologous pair of chromosomes?

   (A) 1
   (B) 2
   (C) 3
   (D) 4
   (E) none of them

33. Consider the fictional animal what-the-hey, that has a diploid number of 16 chromosomes. Approximately how many different combinations of maternal chromosomes are possible in its gametes?

   (A) 32
   (B) 256
   (C) 65,000
   (D) 278,000
   (E) more than 4 million

34. Consider a squirrel who is heterozygous for 3 genes, *GgRrTt*. How many different types of gametes are possible with respect to those genes?

(A) 3

(B) 6

(C) 8

(D) 9

(E) 12

35. A boy whose blood type is O cannot have as his parents

(A) a woman with A and a man with B

(B) a woman with B and a man with B

(C) a woman with B and a man with O

(D) a woman with A and a man with O

(E) a woman with O and a man with AB

36. A man with normal vision married a woman with normal vision, but her father was colorblind. Assume color-vision is a sex-linked trait. What is the probability that the couple will have a daughter who is color-blind?

(A) 0 percent

(B) 12.5 percent

(C) 25 percent

(D) 50 percent

(E) 100 percent

37. Tortoiseshell is a color in cats that is the result of two alleles, one for orange and one for black. The alleles are on the X chromosome. If sex is determined in cats as it is in humans, which statement concerning fur color in cats is true?

(A) Only male cats can be tortoiseshell.

(B) Males need two alleles to have orange fur.

(C) Males need two alleles to have black fur.

(D) Only females can be tortoiseshell under normal conditions.

(E) Tortoiseshell is recessive in males.

38. A certain species of moth may have antennae that are feathery or nonfeathery. Those with feathery antennae usually have red eyes, and those with nonfeathery antennae usually have white eyes. If antenna type and eye color are each controlled by a single pair of genes, a valid conclusion that can be drawn regarding the genes for antenna type and eye color is that these genes are

(A) sex-linked

(B) linked

(C) polyploid

(D) found on the same locus

(E) on separate chromosomes

39. In horses, chestnut coat is dominant to black coat, and trotting gait is dominant to pacing. A horse that was hybrid for both traits would have the genotype *CcTt*. If a horse who is a chestnut pacer is bred several times to a black trotter, and they always have chestnut trotters and black trotters, what are the genotypes of the parents?

   (A) CcTT × ccTT
   (B) CcTt × ccTt
   (C) CCTT × Cctt
   (D) Cctt × ccTT
   (E) Cctt × ccTt

40. Which statement best describes the process of crossing-over?

   (A) It takes place between sister chromatids and results in new gene combinations.
   (B) It takes place between nonhomologous chromosomes and results in increased gene combinations.
   (C) It takes place between homologous chromosomes and results in increased gene combinations.
   (D) It takes place between homologous chromosomes and results in an increased mutation rate.
   (E) It takes place between nonhomologous chromosomes and results in new gene combinations.

41. The correct order of molecules involved in protein synthesis is

   (A) messenger RNA, transfer RNA, DNA, polypeptide
   (B) DNA, messenger RNA, polypeptide, transfer RNA
   (C) polypeptide, DNA, messenger RNA, transfer RNA
   (D) DNA, messenger RNA, transfer RNA, polypeptide
   (E) transfer RNA, polypeptide, messenger RNA, DNA

Questions 42–45 are based on the information below.

In 1973, Stanley Cohen and Herbert Boyer inserted a gene from an African clawed frog into a bacterium. The bacterium then began producing a protein directed by the code found on the inserted frog gene.

42. The newly synthesized genetic material in the bacterium is known as

   (A) recombinant DNA
   (B) messenger RNA
   (C) a favorable gene mutation
   (D) a multiple allele
   (E) a knock-out gene

43. The procedure used by Cohen and Boyer is known as

   (A) cloning
   (B) genetic engineering
   (C) karyotyping
   (D) genetic screening
   (E) restriction analysis

44. Analysis of the DNA from both the frog and the bacterium would reveal that

(A) frog DNA is single stranded, but bacterial DNA is double stranded

(B) frog DNA contains thymine, but bacterial DNA contains uracil

(C) DNA from both organisms is composed of repeating nucleotide fragments

(D) DNA from both organisms contains the sugar ribose

(E) DNA from both organisms contained TATTA boxes

45. Additional copies of the bacterium containing the frog gene could be produced by

(A) asexual reproduction

(B) cross-pollination

(C) inbreeding

(D) grafting

(E) electrical stimulation

Questions 46 and 47 are based on the diagram below, which shows some molecules involved in protein synthesis to answer questions 46 and 47.

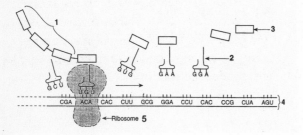

46. Which number indicates part of a molecule containing peptide bonds?

(A) 1

(B) 2

(C) 3

(D) 4

(E) 5

47. The number 2 is pointing to

(A) a tRNA molecule and a DNA triplet

(B) a tRNA molecule and a mRNA codon

(C) a mRNA codon and a DNA triplet

(D) a DNA triplet and an anticodon

(E) a tRNA molecule and an anticodon

48. Alfred Hershey and Martha Chase performed a famous experiment known as the "Blender Experiment." They used both a blender and radioactive sulfur and radioactive phosphorus. Choose the statement among the following that is true.

(A) The radioactive sulfur turned up in the viral progeny.

(B) The blender was used to mix the bacteriophage DNA and the bacterial DNA.

(C) The radioactive phosphorus remained in the protein coat on the outside of the bacteria.

(D) The radioactive phosphorus showed up later in the viral progeny.

(E) There was much less sulfur than phosphorus required to follow the DNA inheritance.

49. Which of the following functions can be attributed to DNA polymerase?

    (A) It replaces RNA nucleotides with DNA nucleotides.

    (B) It fixes errors in the replication of DNA.

    (C) It adds nucleotides to the growing chain at the 3

    (D) It is used in the polymerase chain reaction.

    (E) All of the above are true.

50. Which of the following do prokaryotes use for control of transcription?

    (A) operons

    (B) poly A tails

    (C) guanine caps

    (D) all of the above

    (E) none of the above

51. Transposable elements are an important source of mutations in eukaryotes because they can cause

    (A) inversions

    (B) transpositions

    (C) insertions

    (D) deletions

    (E) all of the above

52. Which of the following gases were probably the **least** and the **most** abundant in the Earth's early atmosphere?

    (A) CO and CO_2

    (B) CO_2 and O_2

    (C) O_2 and CO_2

    (D) O_2 and H_2

    (E) CO and H_2S

53. Which of the following statements about evolution is true?

    (A) Large populations are subject to genetic drift more than small populations.

    (B) Natural selection works on individual phenotypes, not populations.

    (C) Variation in populations arises only through mutations.

    (D) People that lift lots of weights during their reproductive years will have stronger babies than people who do not lift weights.

    (E) None of the above are true.

54. Some young scrub jays in Florida assist parent birds in feeding younger birds at the nest. This is an example of

    (A) altruism

    (B) natural selection

    (C) Lamarkian evolution

    (D) kin selection

    (E) artificial selection

55. The gene for curly fur in squirrels is recessive to the gene for straight fur. If 2 squirrels that had curly fur were put into a park with 6 squirrels that were pure bred for straight hair, how many curly-haired offspring would be expected out of 16 births? Assume that there are equal numbers of male and females and no sexual selection.

    (A) 0

    (B) 1

    (C) 2

    (D) 4

    (E) 6

56. What percentage of the squirrels born in question 55 have the gene for curly hair but do not express it? That is, what is the percentage of carriers?

(A) 6 percent

(B) 18.75 percent

(C) 25 percent

(D) 37.5 percent

(E) 50 percent

Question 57 refers to the following diagrams.

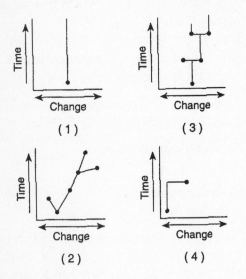

(1)   (3)

(2)   (4)

57. Each point in the graphs above represents a new species. Which graph best represents the concept of gradualism?

(A) 1 only

(B) 2 only

(C) 3 only

(D) 4 only

(E) 3 and 4

58. Darwin's finches, 13 species of finches found in the Galapagos Islands, are good examples of all of the following, **EXCEPT**

(A) adaptive radiation

(B) allopatric speciation

(C) competition as a force driving natural selection

(D) isolating mechanisms

(E) convergent evolution

59. Donald Johanson and others discovered a unique hominid in the early 1970s that they called Lucy. Choose her correct scientific name from the choices below.

(A) *Homo habilis*
    Australopithecus africanus
    Australopithecus afarensis
    Homo erectus
    Australopithecus boisei

60. Which of the following is **NOT** integral to natural selection?

(A) Some individuals are more reproductively fit than others.

(B) Variations can be passed from one generation to the next.

(C) The number of offspring born is related to the amount of resources available.

(D) Individuals who are best suited to their environment will leave more offspring than others.

(E) Adaptations may arise from variations that are inherited.

61. The differences in cell wall structure is used to differentiate

(A) plants from protists

(B) one type of plant from another

(C) one type of prokaryote from another

(D) plants from animals

(E) viruses from protobiotants

62. Which of the following methods or procedures have **NOT** been used to test evolutionary or phylogenetic relationships?

(A) analysis of protein sequences

(B) analysis of DNA sequences

(C) antigen-antibody interactions

(D) phenetic systematics

(E) gel electrophoresis

63. Families are grouped together under what classification level?

(A) order

(B) class

(C) genus

(D) phylum

(E) kingdom

64. The classification levels to which the name *Homo sapiens* belongs are

(A) family and species

(B) order and family

(C) species and genus

(D) genus and species

(E) family and order

65. Which of the following organisms are bilaterally symmetrical?

(A) starfish

(B) sea urchin

(C) coral

(D) jellyfish

(E) none of the above

66. A student on a field trip in Costa Rica found an organism which had the following characteristics: flagella, probably colonial, nucleus present, chitin absent, spores absent, non-photosynthetic. To what kingdom should this organism be assigned?

(A) Animalia

(B) Plantae

(C) Protista

(D) Fungi

(E) Monera

67. Plants have a life cycle that is similar to the algae from which they descended. This life cycle is called the alteration of generations and is illustrated below. Choose the correct position to indicate where meiosis occurs.

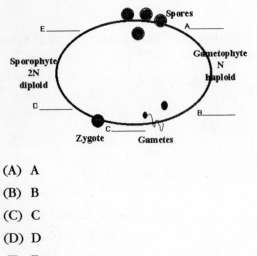

(A) A

(B) B

(C) C

(D) D

(E) E

68. A lichen is an organism that is composed of two separate organisms living symbiotically. The two partners most commonly are

(A) green algae and moss

(B) green algae and fungi

(C) red algae and moss

(D) red algae and fungi

(E) moss and fungi

69. Choose from among the following organisms the one that lacks body cavities and a vascular system.

(A) roundworms

(B) flatworms

(C) rotifers

(D) insects

(E) mollusks

70. Which of the following would be lacking in a moss plant?

(A) xylem and phloem

(B) waxy cuticle

(C) spores

(D) rhizoids

(E) gametophyte tissue

Questions 71–73 are based on the following diagram.

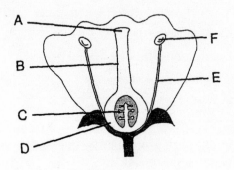

71. The structures that letters E and F are pointing to together represent the

(A) calyx

(B) carpel

(C) sepal

(D) stamen

(E) pistil

72. Both meiosis and fertilization occur in

(A) A and F

(B) B

(C) C

(D) D and E

(E) D and F

320

73. Male gametophytes are produced in

(A) A
(B) F
(C) C
(D) D
(E) E

Questions 74-77 are based on the following diagram.

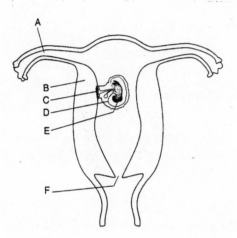

74. The fertilization of the egg that produced the structure E most likely occurred in structure

(A) A
(B) B
(C) C
(D) D
(E) F

75. A structure formed from a combination of maternal and embryonic tissue is located at

(A) A
(B) B
(C) C
(D) D
(E) E

76. Preparation for the implantation of the embryo was regulated by the

(A) cerebellum, pituitary, and testes
(B) hypothalamus, pituitary, and ovaries
(C) uterus, ovaries, and amnion
(D) pituitary, amnion, and ovaries
(E) hypothalamus, pituitary, and amnion

77. The substance within structure D is known as

(A) lymphatic fluid
(B) plasma
(C) yolk
(D) chorionic fluid
(E) amniotic fluid

78. A bluebird reproduces by laying eggs. Which characteristic does the embryo of a bluebird have in common with a human embryo?

(A) implantation in the wall of the uterus
(B) an allantois membrane for protection
(C) a protective shell surrounding the chorion
(D) development within a watery environment inside an amnion
(E) exchange of materials with the mother through a placenta

79. The growth area of a tree that produces the increase in branch length is known as the

(A) xylem
(B) phloem
(C) meristem
(D) pistil
(E) cambium

80. Which of the following are not involved with passage of gases in plants?

    (A) lenticels

    (B) phloem cells

    (C) guard cells

    (D) spongy mesophyll

    (E) stomata

Questions 81–84 are based on the following diagram.

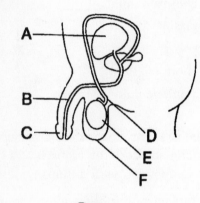

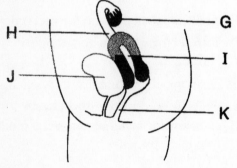

81. Which structures secrete hormones that regulate the development of secondary sexual characteristics?

    (A) A and J

    (B) D and H

    (C) F and I

    (D) E and G

    (E) E and J

82. After sperm are deposited in the female, the pathway they follow to reach the egg is from

    (A) H to I to K

    (B) J to K to H

    (C) G to H to I

    (D) J to K to I

    (E) K to I to H

83. Gametogenesis occurs within structures

    (A) A and J

    (B) A and E

    (C) D and H

    (D) E and J

    (E) E and G

84. Which structures are directly affected by hormones involved in the menstrual cycle?

    (A) G and I

    (B) E and G

    (C) A and J

    (D) I and J

    (E) A and D

85. The function of the counter-current heat exchanger in large predatory fish, like tuna, is to

    (A) trap heat in the muscles

    (B) keep the skin cool in hot water

    (C) rid the fish of too much heat

    (D) warm the blood as it returns to the heart

    (E) cool the water and increase the oxygen content of their gills

86. Earthworms use their nephridia in the same way that people use their

(A) appendixes

(B) cecums

(C) kidneys

(D) bronchi

(E) brains

87. Fats are broken down in the human digestive tract in the _____ by the action of _____.

(A) stomach, pepsin

(B) small intestine, trypsin

(C) stomach, trypsin

(D) small intestine, lipase

(E) stomach, amylase

88. Choose the statement from among the following which is **not** true.

(A) One molecule of cAMP activates one molecule of protein kinase.

(B) cAMP is broken down by adenyl cyclase.

(C) cAMP is involved in the breakdown of glycogen in the liver.

(D) cAMP is formed from ATP.

(E) A hormone acting on a receptor is known as the first messenger, and cAMP is known as a second messenger.

Questions 89–91 are based on the following diagram.

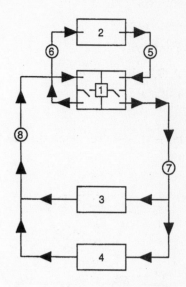

89. Which sequence represents normal blood flow in pulmonary circulation?

(A) $2 \to 5 \to 7 \to 4$

(B) $1 \to 7 \to 3 \to 8$

(C) $1 \to 6 \to 2 \to 5$

(D) $3 \to 8 \to 1 \to 7$

(E) $4 \to 8 \to 1 \to 6$

90. Enlargement and degeneration of tissue within structure 2 will result in

(A) goiter

(B) gout

(C) arthritis

(D) angina

(E) emphysema

91. Number 7 could represent the

(A) dorsal aorta

(B) superior vena cava

(C) inferior vena cava

(D) renal vein

(E) portal vein

92. Voluntary muscle contractions begin with the release of the neurotransmitter _____, followed by _____ ions rushing across the membrane.

(A) acetylcholine, calcium

(B) acetylcholine, sodium

(C) norepinephrine, calcium

(D) epinephrine, calcium

(E) epinephrine, sodium

Question 93 is based on the following diagram.

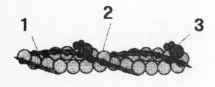

93. The diagram represents the molecules involved with the initiation of a muscle contraction. If 1 represents tropomyosin, then 2 and 3 represent, respectively,

(A) myosin and calcium

(B) myosin and actin

(C) actin and calcium

(D) troponin and calcium

(E) actin and troponin

94. A difference between marsupials and placental mammals is that in marsupials

(A) the young do not receive nourishment from mammary glands

(B) fertilization occurs outside the body of the female

(C) external development occurs without direct nourishment from the mother

(D) the embryo is born at a relatively immature stage of development

(E) the young are always born as twins

95. Which structural features of green plants increase the surface area for absorption of water and minerals?

(A) guard cells

(B) stomata

(C) root hairs

(D) lenticels

(E) companion cells

96. What causes the stem of a tomato plant to bend toward light?

(A) an unequal auxin concentration

(B) a high gibberelin-to-cytokinin ratio in the stem

(C) the transpirational pull of the active lighted side

(D) the unequal concentration of phyto-chrome

(E) the storage of sucrose away from light

97. Double fertilization in plants is a term that denotes

(A) the two sperm nuclei traveling down the style

(B) the union of sperm with two polar nuclei and egg

(C) the two polar nuclei in the ovary that will form the endosperm

(D) the two-celled entrance through the micropyle

(E) most plants can self-fertilize

Questions 98 and 99 are based on the following diagram.

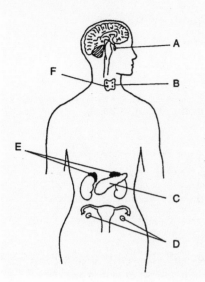

98. A hormone that increases the rate and strength of heart contractions during times of stress is produced in gland

(A) A

(B) B

(C) C

(D) F

(E) E

99. The gland that produces a hormone that regulates the amount of glycogen stored in the liver and muscles is

(A) A

(B) B

(C) C

(D) F

(E) E

Questions 100 and 101 are based on the following diagram.

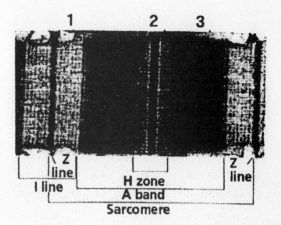

100. Number 2 is situated above the region of the sarcomere consisting of

(A) actin and myosin

(B) myosin and regulatory strands

(C) actin only

(D) myosin only

(E) myofibrils

101. In the sliding filament model of muscle contraction, which one of the following statements is true?

(A) The length of the thin filaments and the length of myosin fibers will not change during a muscle contraction.

(B) During a muscle contraction, the distance between the Z lines will not change.

(C) The actin fibers grab the myosin fibers and pull inwards, reducing the distance between the Z lines.

(D) The size of the sarcomere does not change during a muscle contraction.

(E) All of the above are true.

Questions 102 and 103 are based on the following diagram.

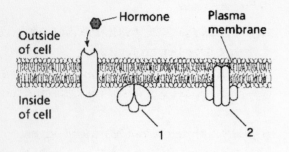

102. The diagram illustrates the principle of

   (A) facilitated diffusion
   (B) signal transduction
   (C) homeostatic feedback
   (D) facilitated transport
   (E) endocytosis

103. In the diagram above,

   (A) 1 could be a relay protein
   (B) 2 could be an effector protein
   (C) 1 could be a G protein that binds to GTP
   (D) 2 could be adenyl cyclase
   (E) all of the above are true

Question 104 is based on the following diagram.

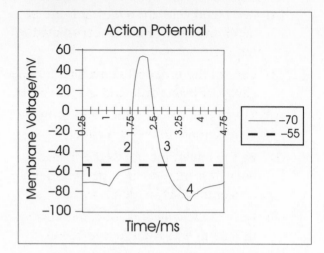

104. Which of the following statements about the action potential is **not** true?

   (A) The magnitude of the action potential is independent of the strength of the stimulus that triggered it.
   (B) Sodium channels have two gates that open or close at different rates.
   (C) Potassium channels have two gates that both operate more slowly than the sodium gates.
   (D) Number 3 depicts a time when the potassium channels are open.
   (E) Number 4 depicts the undershoot that is the result of the sodium channels being closed and the potassium channels being open.

105. Which of the following statements concerning the actions within a nephron is true?

(A) Red blood cells are removed from the nephron at the proximal convoluted tubule.

(B) Cells of the proximal convoluted tubule have many microvilli and mitochondria.

(C) The filtrate entering the Loop of Henle is hypertonic to blood plasma.

(D) By the time urine leaves the human body, it is more than fifteen times more concentrated than blood plasma.

(E) Cells of the collecting duct are impermeable to water but permeable to mineral ions.

106. Which of the following statements about white blood cells is **FALSE**?

(A) They move about using ameboid motion.

(B) They leave the vascular system in search of foreign particles.

(C) They are produced in bone marrow.

(D) They initiate the complement protein reactions that destroy bacteria.

(E) They engulf foreign particles by the act of phagocytosis.

107. Which of the following statements about proteins on our cells known as Major Histocompatability Complex (MHC proteins) are true?

(A) T cells use MHC to rid the body of cells that have been infected.

(B) MHC class II are found on all cells, and MHC class I are found only on immune cells.

(C) MHC proteins coupled with antigenic molecules are the signal needed for destruction by cytotoxic T cells.

(D) A and B are true.

(E) A and C are true.

108. Current treatments for HIV infection include which of the following?

(A) Dideoxynucleotides that keep the virus from replicating

(B) Reverse transciptase inhibitors that bind non-competitively

(C) Soluble CD4 protein that attracts the glycoprotein 41 of the HIV virus

(D) A and B

(E) all of the above

109. Competition between the members of a woodchuck population in a large field could be expected to increase as a result of an increase in the

(A) woodchuck reproduction rate

(B) spread of disease among the woodchucks

(C) number of woodchucks killed by cars

(D) number of secondary consumers

(E) natality rate of their predators

Questions 110 and 111 are based on the following diagram.

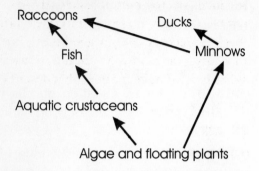

110. Which statement best describes some organisms in the food web above?

   (A) Minnows and fish are primary consumers.

   (B) Algae and floating plants are decomposers.

   (C) Raccoons, fish, and ducks are secondary consumers.

   (D) Aquatic crustaceans are omnivores.

   (E) Algae and floating plants will become saprophytes.

111. Assume the algae and floating plants were evenly distributed in a pond that held 10,000 gallons of water. If there were 500 kilograms of algae and floating plants in the pond, what mass of minnows could one expect to find in any 10 gallons?

   (A) 1 kilogram

   (B) 10 kilograms

   (C) 50 kilograms

   (D) 50 grams

   (E) There is not enough information given to answer the question.

112. Protozoans living in the intestine of a termite secrete enzymes that digest cellulose, providing digestive end products of value to both organisms. This symbiotic relationship could be described as

   (A) mutualism

   (B) commensalism

   (C) parasitism

   (D) saprophytism

   (E) competition

113. The leaves of mistletoe plants photosynthesize, but roots of the mistletoe plant absorb nutrients from living oak trees. The symbiotic relationship is best described as

   (A) mutualism

   (B) commensalism

   (C) parasitism

   (D) saprophytism

   (E) competition

114. Birds that ride on the backs of rhinos and eat the bugs they spot in dung have a relationship with the rhinos that is best described as

   (A) mutualism

   (B) commensalism

   (C) parasitism

   (D) saprophytism

   (E) competition

115. The rain shadow effect is used to describe rainfall around certain mountain ranges. Which of the following statements is true?

   (A) In the Northern hemisphere, there is more rainfall on the eastern slopes.

   (B) The Northern hemisphere and the Southern hemisphere have the same prevailing winds.

   (C) Cool air can hold more moisture than warm air.

   (D) You would expect deserts on the west side of mountain ranges in the Northern hemisphere.

   (E) None of the statements are true.

116. Let r stand for the intrinsic rate of growth, N for the number of organisms in a population, and K for the carrying capacity. A good formula for logistic growth would be

   (A) $rN (N-K)/K$

   (B) $rN (N-r)/rK$

   (C) $rK (K-N)/rN$

   (D) $rK (r-N)/K$

   (E) $rN (K-N)/K$

117. Which of the following biomes is correctly paired with its description?

   (A) tundra—coniferous trees in a cold, dry climate

   (B) savanna—grassland with three distinct climates based mainly on rain

   (C) chaparral—wetlands with many shrubs

   (D) tropical deciduous forests—the most complex communities located along the equator with abundant rainfall

   (E) taiga—cold areas with low, mat-like vegetation

118. Mimicry is an important biotic factor in communities all over the world. For instance, viceroy butterflies closely resemble (mimic) monarch butterflies. Birds that eat viceroys enjoy a good meal, but those that eat monarchs retch right away. What type of mimicry does this describe?

   (A) Mullerian

   (B) aggressive

   (C) crypsis

   (D) Batesian

   (E) deflection

119. Probably the world's most pressing ecological problem is

   (A) destruction of the rain forests

   (B) global warming

   (C) depletion of natural resources

   (D) over-fishing the oceans

   (E) human overpopulation

120. Predators are beneficial to communities because

   (A) they control the prey population growth

   (B) they practice predation harvest

   (C) predators enhance species diversity

   (D) A and B

   (E) A, B, and C

SECTION II

Time–1 Hour and 30 Minutes

Glucose serves as the main unit of currency for the energy needs of humans. How do humans obtain their glucose, what do they do with it, and how do they regulate the amount in their systems?

1. An investigator was working with four traits in tomatoes, two dealing with floral characteristics and two with leaf characteristics. He knew that axial flowers were dominant to terminal flowers and that yellow flowers were dominant to white flowers. He also knew that hairy leaves were dominant to smooth and ovate shape was dominant to oblong. He performed six crosses between dihybrid individuals and their homozygous recessive counterparts. Part of his log follows:

axial, yellow (AaYy) X terminal, white (aayy)
24 axial, yellow
25 axial, white
23 terminal, yellow
28 terminal, white

axial, hairy (AaHh) X terminal smooth (aahh)
43 axial, hairy
8 axial, smooth
7 terminal, hairy
42 terminal, smooth

axial, ovate (AaOo) X terminal, oblong (aaoo)
39 axial, ovate
10 axial, oblong
10 terminal, ovate
41 terminal, oblong

yellow, hairy (YyHh) X white, smooth (yyhh)
26 yellow, hairy
25 yellow, smooth
26 white, hairy
23 white, smooth

yellow, ovate (YyOo) X white, oblong (yyoo)
26 yellow, ovate
24 yellow, oblong
28 white, ovate
22 white, oblong

hairy, ovate (HhOo) X smooth, oblong (hhoo)
48 hairy, ovate
3 hairy, oblong
2 smooth, ovate
47 smooth, oblong

(A) Which, if any, of these genes are linked?

(B) Use the Chi-Squared test to prove your answer to (A) (see Chi-Square table).

(C) Draw a chromosome map representing the above data.

2. Humans have evolved very specific means of dealing with disease-causing organisms. Describe how your body would respond to a bacterial infection, using as many different means of dealing with this bacteria as possible. Do not include non-specific immune responses.

3. A group of students isolated some plant pigments by blending some spinach leaves in a buffer. They attempted to separate the various pigments by both column chromatography and paper chromatography.

(A) Describe the principles behind these types of separatory techniques.

The students ran some whole plant pigments through a spectrophotometer at various wavelengths. They graphed their data and labeled it "whole plant data" (see below).

(B) What colors do you think are represented by 450 nanometers and by 540 nanometers? Why?

Peterson's AP Success: Biology

Critical Points of the Chi Square Distribution
Cumulative probability

	0.005	0.010	0.025	0.05	0.10	0.25	0.50	0.75	0.90	0.95	0.975	0.99	0.995
1	0.39E-4	0.00016	0.00098	0.0039	0.0158	0.102	0.455	1.32	2.71	3.84	5.02	6.63	7.88
2	0.0100	0.0201	0.0506	0.103	0.211	0.575	1.39	2.77	4.61	5.99	7.38	9.21	10.6
3	0.0717	0.115	0.216	0.352	0.584	1.21	2.37	4.11	6.25	7.81	9.35	11.3	12.8
4	0.207	0.297	0.484	0.711	1.06	1.92	3.36	5.39	7.78	9.49	11.1	13.3	14.9
5	0.412	0.554	0.831	1.15	1.61	2.67	4.35	6.63	9.24	11.1	12.8	15.1	16.7
6	0.676	0.872	1.24	1.64	2.20	3.45	5.35	7.84	10.6	12.6	14.4	16.8	18.5
7	0.989	1.24	1.69	2.17	2.83	4.25	6.35	9.04	12.0	14.1	16.0	18.5	20.3
8	1.34	1.65	2.18	2.73	3.49	5.07	7.34	10.2	13.4	15.5	17.5	20.1	22.0
9	1.73	2.09	2.70	3.33	4.17	5.9	8.34	11.4	14.7	16.9	19.0	21.7	23.6
10	2.16	2.56	3.25	3.94	4.87	6.74	9.34	12.5	16.0	18.3	20.5	23.2	25.2
11	2.60	3.05	3.82	4.57	5.58	7.58	10.3	13.7	17.3	19.7	21.9	24.7	26.8
12	3.07	3.57	4.40	5.23	6.30	8.44	11.3	14.8	18.5	21.0	23.3	26.2	28.3
13	3.57	4.11	5.01	5.89	7.04	9.3	12.3	16.0	19.8	22.4	24.7	27.7	29.8
14	4.07	4.66	5.63	6.57	7.79	10.2	13.3	17.1	21.1	23.7	26.1	29.1	31.3
15	4.60	5.23	6.26	7.26	8.55	11.0	14.3	18.2	22.3	25.0	27.5	30.6	32.8
16	5.14	5.81	6.91	7.96	9.31	11.9	15.3	19.4	23.5	26.3	28.8	32.0	34.3
17	5.70	6.41	7.56	8.67	10.1	12.8	16.3	20.5	24.8	27.6	30.2	33.4	35.7
18	6.26	7.01	8.23	9.39	10.9	13.7	17.3	21.6	26.0	28.9	31.5	34.8	37.2
19	6.84	7.63	8.91	10.1	11.7	14.6	18.3	22.7	27.2	30.1	32.9	36.2	38.6
20	7.43	8.26	9.59	10.9	12.4	15.5	19.3	23.8	28.4	31.4	34.2	37.6	40.0
21	8.03	8.90	10.3	11.6	13.2	16.3	20.3	24.9	29.6	32.7	35.5	38.9	41.4
22	8.64	9.54	11.0	12.3	14.0	17.2	21.3	26.0	30.8	33.9	36.8	40.3	42.8
23	9.26	10.2	11.7	13.1	14.8	18.1	22.3	27.1	32.0	35.2	38.1	41.6	44.2
24	9.89	10.9	12.4	13.8	15.7	19.0	23.3	28.2	33.2	36.4	39.4	43.0	45.6
25	10.5	11.5	13.1	14.6	16.5	19.9	24.3	29.3	34.4	37.7	40.6	44.3	46.9
26	11.2	12.2	13.8	15.4	17.3	20.8	25.3	30.4	35.6	38.9	41.9	45.6	48.3
27	11.8	12.9	14.6	16.2	18.1	21.7	26.3	31.5	36.7	40.1	43.2	47.0	49.6
28	12.5	13.6	15.3	16.9	18.9	22.7	27.3	32.6	37.9	41.3	44.5	48.3	51.0
29	13.1	14.3	16.0	17.7	19.8	23.6	28.3	33.7	39.1	42.6	45.7	49.6	52.3
30	13.8	15.0	16.8	18.5	20.6	24.5	29.3	34.8	40.3	43.8	47.0	50.9	53.7
31	14.5	15.7	17.5	19.3	21.4	25.4	30.3	35.9	41.4	45.0	48.2	52.2	55.0
32	15.1	16.4	18.3	20.1	22.3	26.3	31.3	37.0	42.6	46.2	49.5	53.5	56.3
33	15.8	17.1	19.0	20.9	23.1	27.2	32.3	38.1	43.7	47.4	50.7	54.8	57.6
34	16.5	17.8	19.8	21.7	24.0	28.1	33.3	39.1	44.9	48.6	52.0	56.1	59.0
35	17.2	18.5	20.6	22.5	24.8	29.1	34.3	40.2	46.1	49.8	53.2	57.3	60.3
36	17.9	19.2	21.3	23.3	25.6	30.0	35.3	41.3	47.2	51.0	54.4	58.6	61.6
37	18.6	20.0	22.1	24.1	26.5	30.9	36.3	42.4	48.4	52.2	55.7	59.9	62.9
38	19.3	20.7	22.9	24.9	27.3	31.8	37.3	43.5	49.5	53.4	56.9	61.2	64.2
39	20.0	21.4	23.7	25.7	28.2	32.7	38.3	44.5	50.7	54.6	58.1	62.4	65.5
40	20.7	22.2	24.4	26.5	29.1	33.7	39.3	45.6	51.8	55.8	59.3	63.7	66.8
41	21.4	22.9	25.2	27.3	29.9	34.6	40.3	46.7	52.9	56.9	60.6	65.0	68.1
42	22.1	23.7	26.0	28.1	30.8	35.5	41.3	47.8	54.1	58.1	61.8	66.2	69.3
43	22.9	24.4	26.8	29.0	31.6	36.4	42.3	48.8	55.2	59.3	63.0	67.5	70.6
44	23.6	25.1	27.6	29.8	32.5	37.4	43.3	49.9	56.4	60.5	64.2	68.7	71.9
45	24.3	25.9	28.4	30.6	33.4	38.3	44.3	51.0	57.5	61.7	65.4	70.0	73.2
D.F.	0.005	0.010	0.025	0.05	0.10	0.25	0.50	0.75	0.90	0.95	0.975	0.99	0.995

During the separation, the students with the paper chromatography cut out the lines that looked alike and pooled them into dishes.

(C) What solution should they use to elute the pigments from the paper?

During the separation, the students with the column chromatography collected three different fractions by holding three different test tubes under the column over a 10 minute period. They then ran each through a spectrophotometer, graphed the data, and labeled the fractions 1, 2, and 3.

(D) Compare fractions 1 and 2. Did the students separate different pigments? Why or why not?

(E) Compare fraction 3 to fraction 1.

(F) What substance do you think is in the whole plant and in fraction 3 but is not in fraction 1 and fraction 2?

ANSWERS AND EXPLANATIONS

SECTION 1

1. **The correct answer is (E).** Cohesion is the attraction of water molecules for one another, which is defined in the first three choices.

2. **The correct answer is (B).** Blood is made up of cells and plasma, the latter accounting for about 55 percent of the total by volume. Since plasma is around 90 percent water, about 50 percent (0.9 X 55%).

3. **The correct answer is (D).** 0.25 liters is 250 milliliters. 171 grams in 0.5 liters would be 1 molar (342/1=171/0.5). If added to only half that amount, the resulting molarity would be twice as great, or 2.0.

4. **The correct answer is (C).** pH is a logarithmic scale, so that an increase of 4 on the scale is 10^4, or 10000. It cannot be 1/10000 because pH 4 has many more hydrogen ions than pH 8.

5. **The correct answer is (A).** Choice (A) represents a glucose molecule, the primary energy source for animals.

6. **The correct answer is (B).** A lipid is synthesized from fatty acids, represented by the letter E, and glycerol, represented by the letter B.

7. **The correct answer is (D).** Amino acids are composed of 4 parts: the amino group (NH_2), the central carbon (CH), the variable group (R, or in this case, the methyl group CH_3), and the carboxyl group (COOH).

8. **The correct answer is (B).** The table is mRNA codons, therefore the DNA that would be complementary to C-G-A would be G-C-T, using Watson-Crick base pairing rules (A with T and G with C).

9. **The correct answer is (A).** Phenylalanine is the amino acid attached to the tRNA molecule with AAA as its anticodon. Using Watson-Crick

base pairing, U bonds with A. Because anticodons are complementary to mRNA codons, which are complementary to DNA triplets, the anticodons are essentially the DNA triplet sequence, with the exception that uracil (U) is the RNA base that substitutes for thymine (T).

10. **The correct answer is (B).** The peptide bonds are between the amino acids, each of which have a variable group, R.

11. **The correct answer is (C).** The polypeptide is 3 amino acids joined together. Amino acids are composed of 4 parts: the amino group (NH_2), the central carbon (CH), the variable group (R), and the carboxyl group (COOH).

12. **The correct answer is (E).** Entropy is a measure of orderliness or disorder. Disorder increases as the temperature increases.

13. **The correct answer is (A).** While most enzymatic reactions are very rapid, and each has its own rate for the conversion of substrate, choice (A) reflects the idea that there are more substrate molecules than can be handled by the number of enzyme molecules. When the available enzyme molecules become saturated, the rate of the reaction plateaus.

14. **The correct answer is (D).** All fermentation is less efficient than cellular respiration, resulting in 2 ATP rather than the 34–36 that can be expected with full cellular respiration. Alcoholic fermentation results in the formation of alcohol (such as wine and spirits) and carbon dioxide, which bubbles off.

15. **The correct answer is (D).** Keq is equal to the concentration of the products over the concentration of the reactants at chemical equilibrium. If Keq is less than 1, then the reverse reaction dominates; if Keq is greater than 1, the forward reaction dominates. If the free energy of the products is greater than the free energy of the reactants, then energy will be required to form the products, and the reaction will be endergonic.

16. **The correct answer is (A).** Glycolysis not only yields ATP but gets NAD^+ reduced to NADH. Two pyruvate molecules are also produced.

17. **The correct answer is (B).** Chemiosmosis, the production of a hydrogen ion gradient used to produce ATP, is done in both mitochondria and chloroplasts. The hydrogen ions diffuse through ATP synthase from the intermembrane space to the matrix of the mitochondria and from the thylakoid compartment to the stroma in the chloroplasts.

18. **The correct answer is (C).** NADPH has a role in photosynthesis, not cellular respiration, and there are no dehydrogenases in the respiratory chain.

19. **The correct answer is (D).** Vesicles carry products of the ER to the golgi apparatus as well as products of the Golgi to the plasma membrane.

20. **The correct answer is (C).** Students should be able to recognize organelles by appearance and by function.

21. **The correct answer is (B).** The most important intercellular junction in plants is the plasmodesma, the space between the cell walls through which cytoplasm passes freely.

22. **The correct answer is (C).** Both mitochondria and chloroplasts, which have their own prokaryotic-like DNA, are candidates for this important symbiotic relationship.

23. **The correct answer is (D).** The ultracentrifuge has been used for years, not only to isolate organelles but DNA as well.

24. **The correct answer is (E).** PEP stands for phosphoenolpyruvate, and the PEPcarboxylase is the enzyme that is used to assimilate carbon dioxide. Rubisco is used in the Calvin cycle in the bundle sheath cells, which serve to isolate rubisco from the atmosphere.

25. **The correct answer is (C).** In the light reactions or photophosphorylation, water is broken down and oxygen is given off as a waste product. NADP and ADP are reduced and phophorylated, respectively, by the energy of the Sun.

26. **The correct answer is (D).** Water potential is influenced by the amount of solutes in the solution. Pure water has a water potential value of 1, and when solutes are added, the value becomes negative. The left side initially has fewer solutes and therefore a higher water potential. Water always flows from a higher to a lower water potential, and, therefore, the water level will rise on the right side.

27. **The correct answer is (B).** Because the membrane is semipermeable, glucose will not pass through the membrane, but water and NaCl will. Water will move to a lower water potential, and NaCl will diffuse from left to right (since diffusion is the movement of molecules from a higher to a lower concentration).

28. **The correct answer is (A).** Plants that open their stomata only at night belong to the Crassulacean Acid Metabolism (CAM) group.

29. **The correct answer is (B).** The levels of cyclin fluctuate during the cell cycle, but the amount of cdks is relatively constant.

30. **The correct answer is (E).** The number of centromeres per chromosome is always one. At prophase, the chromosomes have duplicated, and two chromatids make up a chromosome.

31. **The correct answer is (C).** Meiosis reduces the chromosome number by 2 cell divisions, so that one cell divides into 2 cells, and they each divide, giving 4 cells. Since the chromosomes doubled at the beginning of meiosis, each of these 4 cells receives 23, or the haploid number of chromosomes.

32. **The correct answer is (B).** Homologous chromosomes have to be the same size, and they have the same information (genes) at the same loci (position), represented by the horizontal lines in the drawing. There can be, of course, different information, or alleles, within each pair of genes.

33. **The correct answer is (B).** The number of combinations is calculated by the equation $n = 2^y$, where y is equal to the number of chromosomes in the gamete, which is the haploid number 8 in this case.

34. **The correct answer is (C).** The equation above can be used or an alternative method would be that *GgRg* can produce 4 types with *T* and another 4 types with *t*.

35. **The correct answer is (E).** Both A and B alleles are dominant to O. The man with AB will contribute either the A or the B to his child and therefore cannot have an O child.

36. **The correct answer is (A).** Since the man will contribute his X to the daughter, she will receive his gene for color vision and cannot be colorblind. Half of the couples' boys have the chance of being colorblind.

37. **The correct answer is (D).** Since females have two X chromosomes, only they have the ability to have both an orange and a black allele.

38. **The correct answer is (B).** Genes that do not assort independently are linked. They are not assumed to be sex-linked in this example because no ratios were given that separated male and female differences.

39. **The correct answer is (D).** Because they always have trotters and one parent is a pacer, the other parent must be homozygous dominant (tt Î TT). They have half chestnut and half black because the chestnut parent is hybrid and the black, of course, is homozygous recessive (Cc Î cc).

40. **The correct answer is (C).** Crossing-over between sister chromatids would be indeterminable since they are identical. Only homologous chromosomes cross-over, exchanging genetic material, because only they form tetrads or bivalents.

41. **The correct answer is (D).** The central dogma of modern biology is that DNA is transcribed to RNA, which is translated into protein.

42. **The correct answer is (A).** Recombinant DNA is the result of splicing DNA from one source and placing it into another (factory) source by the use of restriction enzymes.

43. **The correct answer is (B).** Genetic engineering uses the tools of restriction enzymes to cut DNAs and ligases to splice them together.

44. **The correct answer is (C).** All DNA is made up of nucleotides, which have as their constituent parts a phosphate group, a nitrogen base, and a deoxyribose sugar.

45. **The correct answer is (A).** Genetic engineering depends on the successful and rapid reproduction of the recombinant DNA that asexual reproduction provides. While bacteria have been the mainstay for genetic engineering, even mammals are being asexually reproduced (cloned) to give the same recombinant DNA products for which they were designed.

46. **The correct answer is (A).** The polypeptide being produced is polymerization of amino acids joined by peptide bonds.

47. **The correct answer is (E).** The anticodon is the 3 bases that will form complementary base pairs with the mRNA. The molecule is a tRNA (transfer RNA) molecule.

48. **The correct answer is (D).** Hershey and Chase demonstrated in the blender experiment that DNA, which contains phosphorus, not sulfur, was the hereditary material, not proteins, which contain sulfur but not phosphorus.

49. **The correct answer is (E).** DNA polymerase exists in three different forms in eukaryotic cells and performs all those functions as well as adding nucleotides to the lagging strand.

50. **The correct answer is (A).** The other two choices are eukaryotic only.

51. **The correct answer is (E).** Most transposable elements, first discovered by Barbara McClintock in maize, require an RNA intermediate in eukaryotes, unlike prokaryotes. They can change the function or expression of genes and perhaps create new versions of existing genes.

52. **The correct answer is (C).** Oxygen was rare, in part because there were no producers in the early history of Earth. Carbon dioxide was plentiful, probably setting up a greenhouse effect that lasted for a geologically long time.

53. **The correct answer is (B).** Natural selection works on individuals, and evolution works on populations. Small populations are much more subject to genetic drift than large populations. The idea about weights is patently false—Lamarkian through and through.

54. **The correct answer is (D).** Kin selection increases the fitness of the individual, while altruism (which it appears to be) would lessen individual fitness.

55. **The correct answer is (B).** Calculate the gene frequency by finding the proportion of curly genes: 2 squirrels with 2 genes each is 4 genes for curly hair. The total number of genes in the population is 16 since there are 8 squirrels with 2 genes each. Therefore, $4 \div 16$ is 0.25, the frequency for s, the curly gene. All squirrels with curly hair in the next generation will be ss, and the genotypic frequency will be 0.25×0.25 or 0.0625. Since there are 16 individuals born the next generation, $16 \times 0.0625 = 1$.

56. **The correct answer is (D).** Again, the gene frequency is required. We know from above that the recessive allele(s) has a frequency of 0.25. The Hardy–Weinberg theorem states that the frequency of both alleles must sum to 1, therefore, $S = 0.75$. The frequency of the carriers is $S \times s$ and $s \times S$, or $2 \times S \times s$ (2pq in HW lingo). To change any number to a percentage, multiply by 100%.

57. **The correct answer is (B).** Darwin's theory of gradualism as the speed of evolution is illustrated as ongoing change over time—slow and steady. Gould and Eldridge proposed that punctuated equilibrium was more likely a true reflection, one that can be illustrated by either 3 or 4. The pattern here is one of little or no change over long periods and then rapid bursts of change, usually caused by some catastrophic event.

58. **The correct answer is (E).** Convergent evolution describes the similar appearance of species that do not share a recent common ancestor. Adaptive radiation, on the other hand, describes the dissimilarity between species that do share such a relative. Adaptive radiation is the radiating out into new niches and the eventual isolation that follows.

59. **The correct answer is (C).** *A. africanus* was described by Raymond Dart in the 1920s, the first australopithecine found.

60. **The correct answer is (C).** The amount of resources in the environment will never be sufficient to feed the reproductive possibilities that exist. Darwin garnered this idea from his observations and by reading Malthus's *Dismal Theorem*.

61. **The correct answer is (C).** Eubacteria have long been differentiated by the Gram-stain technique, which capitalizes on the amount and location of peptidoglycan in the cell wall. Archaebacteria, also prokaryotic, have been shown to have a cell wall lacking in peptidoglycan.

62. **The correct answer is (D).** Phenetic systematics does not interest itself with evolutionary relationships but compares only structures present or absent among different organisms.

63. **The correct answer is (A).** The largest classification level is the domain, followed by kingdom, phylum, class, order, family, genus, and species.

64. **The correct answer is (D).** *Homo sapiens* is the scientific name for our genus and species and has been translated as wise or knowing (*sapiens*) man (*Homo*).

65. **The correct answer is (E).** All of the organisms listed belong to two important marine phyla, the Echinodermata and Cnidaria, most members of which are radially symmetrical.

66. **The correct answer is (C).** The organism cannot be a Moneran because it has a nucleus. It cannot be a fungus because it lacks chitin and does not produce spores. It cannot be a plant because it is non-photosynthetic. Therefore, in the true nature of the kingdom, by process of elimination, it is a Protist.

67. **The correct answer is (E).** Meiosis serves to reduce the chromosome number from 2N to N.

68. **The correct answer is (B).** Lichens are classified as fungi, mostly because the outer part of the organism is a fungus. While some cyano-bacteria are their symbiotic partners, most of the time it is a small green algae.

69. **The correct answer is (B).** Animals that lack body cavities are acoelomatic.

70. **The correct answer is (A).** Xylem and phloem are the two major types of vascular tissue. Mosses lack vascular tissue and, as such, are limited in their size.

71. **The correct answer is (D).** The stamen is composed of the filament (E) and the anther (F).

72. **The correct answer is (C).** The ovule is where eggs are produced by meiosis and where the sperm reach the egg after traveling through the pollen tube.

73. **The correct answer is (B).** Pollen grains, with their sperm, are produced in the anther (F).

74. **The correct answer is (A).** The oviducts or fallopian tubes are where the sperm usually meet the egg, which then descends. If it does not, a tubal pregnancy may develop.

75. **The correct answer is (C).** The placenta is formed jointly between the embryo and the endometrium of the mother.

76. **The correct answer is (B).** The hypothalamus regulates the level of FSH and LH released by the pituitary, which influences the production of estrogen and progesterone by the ovaries.

77. **The correct answer is (E).** The amnion is within the chorion, which is the outermost membrane.

78. **The correct answer is (D).** The allantois is used by developing chicks to store uric acid and for respiration. There is neither placenta nor uterus involved.

79. **The correct answer is (C).** Meristematic tissue is responsible for apical or tip growth, both in the roots and the shoots. Cambium produces xylem and phloem, which give rise to lateral growth.

80. **The correct answer is (B).** Phloem cells carry dissolved sugars and are vascular tissue. Spongy mesophyll cells are found in the lower half of leaves, where most gas exchange takes place. Lenticels are found on the stems.

81. **The correct answer is (D).** The testes and ovaries produce the androgens and estrogens that influence secondary sexual characteristics, like deeper voices and breast development.

82. **The correct answer is (E).** Sperm travel up through the vagina, through the cervix, into the uterus, and usually fertilize the ovum in the oviduct.

83. **The correct answer is (E).** Gametogenesis is the production of gametes by meiosis, which takes place in the testes and ovaries.

84. **The correct answer is (A).** The menstrual cycle occurs only in women. Both the ovaries and the uterus are target tissues of different hormones such as FSH, LH, estrogen, and progesterone.

85. **The correct answer is (D).** Tuna use the heat that is generated by powerful swimming muscles to heat the blood, which has a circulatory pattern, unlike fish that lack this adaptation.

86. **The correct answer is (C).** Nephridia are one of the organs repeated in each segment of the earthworm. They are their chief excretory organs.

87. **The correct answer is (D).** The stomach serves to digest proteins. The small intestine first uses bile that is delivered from the gall bladder to break down fat droplets into smaller droplets, which present a larger surface area on which the lipase can work.

88. **The correct answer is (B).** Adenyl cyclase is the enzyme that converts ATP to cAMP.

89. **The correct answer is (C).** The right side of the heart serves the lungs (pulmonary system). It is also the shortest of the two systems.

90. **The correct answer is (E).** Emphysema is the loss of the alveolar sacs that provide the large surface area for the exchange of gases in the lungs.

91. **The correct answer is (A).** Choices (B) through (E) are all associated with veins, which take blood to, not away, from the heart.

92. **The correct answer is (B).** Acetylcholine begins a muscle contraction when it is released from the synaptic vesicles. While calcium is required for a contraction, it is sodium that causes the release of the calcium.

93. **The correct answer is (E).** Tropomyosin covers up the binding site to which myosin heads are attracted. When calcium (not pictured) and troponin bond, it changes the configuration of troponin, which moves tropomyosin, revealing the binding sites.

94. **The correct answer is (D).** External development for later stages is a hallmark of these pouched mammals.

95. **The correct answer is (C).** Most water comes from the roots, and the fine branching of these hairs is another example of the importance of increasing the surface area for many organs.

96. **The correct answer is (A).** Plant growth toward light (positive phototropism) is caused by auxin's effect on cell elongation. The cells on the darker side of the plant stem are growing faster than the ones on the lighter side. F. W. Went demonstrated that auxin-enriched agar blocks placed offset on plant tips could cause this bent growth, even in the dark.

97. **The correct answer is (B).** One sperm fertilizes the two polar nuclei to form endosperm, and another sperm fertilizes the egg to create the embryo.

98. **The correct answer is (E).** The adrenal medulla (the inner part of the adrenal glands) produces both epinephrine and norepinephrine, which function in that capacity.

99. **The correct answer is (C).** The Islets of Langerhans located in the pancreas (C) secrete glucagon and insulin.

100. **The correct answer is (D).** The area under 1 is thin filaments only, and that under 2 is myosin only. Thick (myosin) and thin filaments are found under number 3.

101. **The correct answer is (A).** The fibers themselves do not change size. Rather, the myosin heads attach to the actin and pull inward, reducing the size of the I band or I line.

102. **The correct answer is (B).** Signal transduction is passing information to the cell, and having it amplified within the cell, to direct the cell's activities. An example would be the hormone glucagon landing on its receptor sites in the liver to cause the conversion of glycogen to glucose.

103. **The correct answer is (E).** In signal transduction, the receptor-hormone complex activates a stimulatory G protein that binds with GTP, causing a conformation change in the effector protein, which could be adenyl cyclase, which converts ATP to the second messenger cAMP.

104. **The correct answer is (C).** Potassium channels have only one gate while sodium channels have two gates.

105. **The correct answer is (B).** Red blood cells never enter the nephron because the capillaries of the glomerulus are not permeable to such large particles. The filtrate entering the Loop of Henle is isotonic to blood plasma, and only desert mammals like kangaroo rats or bats are as efficient at conserving water, as noted in choice (D). Human urine is about four times as concentrated as blood plasma. The cells of the collecting duct are permeable to water, which undergoes osmosis to the interstitial fluid.

106. **The correct answer is (D).** Antibodies are the signal for complement attack.

107. **The correct answer is (E).** MHC class I are found on all nucleated mammalian cells.

108. **The correct answer is (D).** Soluble CD4 was tested early and found to attract gp 120, the viral protein that allows entry into the cell. It is not a current treatment.

109. **The correct answer is (A).** Intraspecific competition will increase with the greater number of woodchucks as resources will become limited. All other choices reduce the woodchuck population.

110. **The correct answer is (C).** Producers (green plants and algae) are eaten by primary consumers, which are eaten by secondary consumers.

111. **The correct answer is (D).** Primary consumers can be expected to capture 10 percent of the energy and biomass of the trophic level below them. Therefore, there will be 50 kg. of minnows in 10,000 gallons, or 0.050 kg. in 10 gallons, which is 50 grams in 10 gallons.

112. **The correct answer is (A).** Mutualism describes a relationship that is beneficial for both parties.

113. **The correct answer is (C).** Parasitism describes a relationship that benefits one party to the detriment of the other.

114. **The correct answer is (B).** Commensalism benefits one party, but the other is not affected.

115. **The correct answer is (E).** The prevailing winds blow west to east, bringing moist air. When a mountain range is encountered, the air rises and cools but can hold less moisture. After the rain, there is less moisture for the east side of the mountain.

116. **The correct answer is (E).** As N approaches K, the growth rate will decrease. Substitute values for K in the expression (K-N)/K in order to see its effect on rN.

117. **The correct answer is (B).** The tundra and taiga were switched and substituted "rain" for "deciduous" in choice (D). The chaparral is a dry climate, subject to periodic burns.

118. **The correct answer is (D).** All mimicry relationships have three parties: the model, the mimic, and the dupe. The monarch is the model, the viceroy is the mimic, and the predator is the signal receiver or dupe. This is true of all Batesian mimicries.

119. **The correct answer is (E).** Overpopulation indirectly is the cause of all the other problems.

120. **The correct answer is (E).** Predators enhance species diversity by affecting the most numerous prey species in a community, which are frequently superior competitors. Predation serves to lessen the competitive exclusion principle. The predation harvest is helpful because predators remove sick, older, and less reproductively fit prey (which unfortunately can also be the young; but then, they are reducing the prey numbers and practicing prey control—keeping prey species from eating themselves out of house and home).

SECTION II

1a. Humans obtain glucose from

- digestion of plants, animals, other organic substances
- hydrolysis of glycogen
- gluconeogenesis (the production of glucose from non-carbohydrate substances, like muscle proteins)

1b. Humans use glucose for

- glycolysis and production of ATP
- cellular respiration and production of ATP
- carbon skeleton for structural molecules, like polysaccharides
- polymerization of energy storage molecules, like glycogen

1c. Humans regulate levels of glucose with hormones

- insulin reduces blood glucose by having it taken up by cells
- glucagon increases blood glucose by hydolysis of glycogen, gluconeogenesis, and secretion of glucose by cells

- epinephrine and norepinephrine cause glucose levels to rise from glycogen conversion in liver and muscles and glucose release by liver

2a. Linked genes

- axial, hairy, and ovate
- yellow is only trait with 1/4:1/4:1/4:1/4 ratio in crosses
- above ratio is expected with genes that assort independently
- genes that do not assort independently are linked
- crossing-over results in non-Mendelian ratios of other crosses
- crossing-over produces recombinant chromosomes
- recombinant chromosomes are recognized as non-parental
- non-parental types, like axial, smooth in second cross, are result of recombination

2b. Typical Chi-square calculation

- $\chi^2 = \Sigma \, (obs\text{-}exp)^2/exp$
- null hypothesis: axial and yellow are linked genes
- assume: independent assortment with equal numbers of each offspring (1/4 each)

Observed	Expected	(O-E)	$(O\text{-}E)^2$	$(O\text{-}E)^2/E$
24	25	1	1	0.04
25	25	0	0	0
23	25	2	4	0.16
28	25	3	9	0.36

Chi-squared $(\chi^2) = 0.56$

Chi-squared is less than 7.8, the table value for 95% significance at 3 degrees of freedom

Observed and expected are statistically the same.

Conclusion: axial and yellow assort independently and are therefore not linked.

(Sample calculation to demonstrate linkage: axial and hairy)

Observed	Expected	(O-E)	$(O\text{-}E)^2$	$(O\text{-}E)^2/E$
43	25	18	324	12.96
7	25	18	324	12.96
8	25	17	289	11.56
42	25	17	289	11.56

$\chi^2 = 49.04$

Chi-squared is more than 7.8, the table value for 95% significance at 3 degrees of freedom

Observed and expected are statistically different

Conclusion: axial and hairy don't assort independently and are therefore linked.

2c. Chromosome map

Let homologous chromosomes be represented by

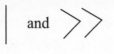 and

Then,

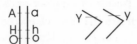

depicts the map of the chromosomes. Furthermore, from the above data, the distance in centimorgans or map units between the genes can be calculated by finding the percentage of recombinants. For example, in the axial and hairy cross: $(8+7)/100 \times 100\% = 15\%$ recombinants and 15 map units between axial and hairy.

The first chromosome would then be illustrated

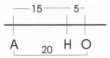

3. Specific response to bacteria

(a) Humoral response (not cell mediated)

- A few B cells have receptors that will recognize (attach to) bacterial site (known as antigenic determinant)
- receptors when floating free in blood and lymph are known as antibodies
- antibodies
- may keep bacteria from adhering to surfaces
- may encourage phagocytosis by macrophages
- may initiate destruction by complement proteins
- specific B cells that recognize the bacteria will be cloned (clonal selection theory)
- activated B cells will differentiate into
- plasma cells that produce antibodies
- memory cells that will recognize future bacterial infection of the same type

(b) Cell mediated response

- macrophge ingestion of bacteria
- presentation of processed antigen bound to Class II MHC protein
- helper T cell responds to antigen and MHC by proliferating
- some helper T cells become memory T cells
- helper T cells activate B cells with lymphokines
- B cells proliferate and differentiate into
- plasma cells
- memory cells

- cytotoxic T cell may kill cells infected with bacteria
- cytotoxic T cell recognizes Class I MHC protein and antigenic determinant
- cytotoxic T cells proliferate and differentiate into
- Tc (cytotoxic T) and Tc memory cells
- cytotoxic T cells destroy infected cells with perforin

4. Plant pigment lab answers

(a) The principle for chromatography is that the pigments will travel at different rates due to

- solubility in the solvent
- attaction for the paper
- attraction for the matrix
- size of the pigment molecule (while the smallest molecules move fastest on paper, they often move slower in a column because they enter the molecules of the matrix or move in a flowing pattern around the matrix molecules)

(b) The color at 450 is blue and absorption is high.

- The color at 540 is green because the absorption is lowest and plants reflect, but do not absorb, green light (without accessory pigments).

(c) The solution should be the same as that in which it was dissolved.

- Most students have used acetone-petroleum ether mixture.

(d) Fractions 1 and 2 are the same solution because the peaks fall at the same wavelengths—fraction 2 is half the concentration.

- The amplitude reflects strength or concentration.

(e) Fraction 3 is obviously a different solution.

- Fraction 3 has a maximum absorption in the 500 nm range.

(f) The substance that is in both the whole plant and fraction 3 is an accessory pigment.

- The accessory pigment has to be in the whole plant solution since fraction 3 was derived from the whole plant solution.

NOTES

NOTES

Peterson's is on your side with everything you need to get ready for college!

Get on line at petersons.com for a jump start on your college search.

Search our college database
Get financial aid tips
Browse our bookstore

And when you're ready to apply, you're ready for CollegeQuest.com!

CollegeQuest.com is our **free** online college application service that lets you apply to *more colleges than anyone else on the Internet!*

Why CollegeQuest.com?
Fill out one application for 1,200 colleges!
Talk with admissions deans!
It's FREE!

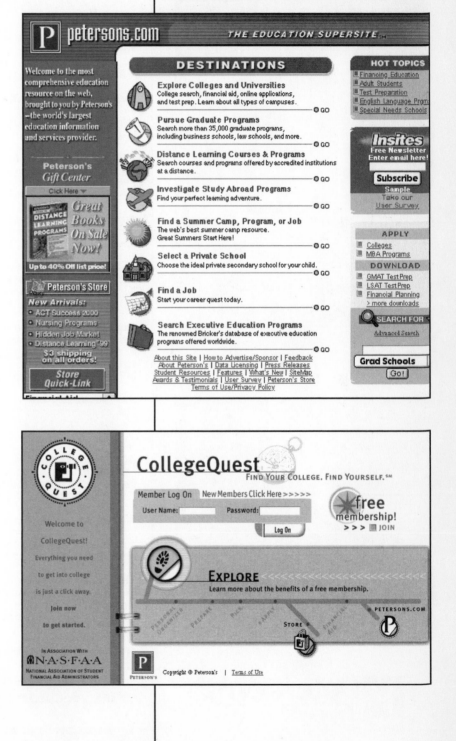

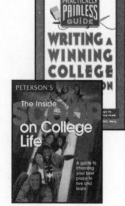